CONTENTS

Preface
A. Markoff

There is certainly more than one way to compose a book about handling, interpreting and applying the information contained in the DNA molecule of living organisms. Not necessarily delving into trivial characterizations such as "molecule of life", or reducing every quantifiable feature of DNA to "information content", this book is trying to deliver the view (and the tools) of a systems biology perspective. On the background of the nearly finished human genome sequence, views could be polarized regarding the necessity of bioinformatic tools and their application for solving biological questions. Among those, who simplistically believe that "knowing the genes" is knowing the function, this book could at least seed the grain of curiosity and wish to understand how actually structural freedoms (and constraints) on DNA, genes and genomes levels contribute to what we call "biological complexity". To those, who already look for tools to solve such complexity issues, available Internet resources offer a wealth of analytical options and solution algorithms. Shopping for such tools is a straightforward and on many occasions unburdened process, if one is knowing what to look for exactly. At times however simple labels are not enough to mark the diversifying richness of available "departmental store" supplies or the user simply does not know what they mean. This is why some sort of a guide for the novice (and experienced) users is valuable to bring pointers and organizing power into the wealth of sequence data. This book is intended as a useful tool for academic and industrial or private researchers in the field of life sciences, or those concerned with genetic information.

Genetic information in living systems is only meaningful in the context of developmental change and acquired functional diversity. This is why the book starts with an algorithmic approach of phylogeny estimation from available sequence data. It proceeds then in the next chapter with the process of defining and classifying DNA sequence information in discrete units of heredity, the process known as "gene annotation". A wealth of data on gene expression experiments is available for the user and the next chapter deals with ways to filter, organize and interpret this information. Further, the reader is guided "directly to the source" of the most powerful and useful molecular biology industrial power "open source" package EMBOSS, which has come to replace GCG for sequence analysis. Unfortunately, at the time of publishing this book, the UK Medical Research Council has decided to close the Research and Bioinformatics Divisions of the Rosalind Franklin Centre for Genomic Research (the current home of EMBOSS).

A chapter describing integrated bioinformatical approaches to gene family studies is a core highlight of this book, which demonstrates how empirical research combines with selectively applied analytical tools to elucidate (patho)physiological function.

Next, an alternative and less powerful, but definitely useful possibility to deal with sequence information in the context of "local" gene projects is described in the "desktop sequence analysis" tools chapter. Following chapters deal with discrete DNA elements with "meaningless" nature, s.c. repetitive elements, and elements with regulatory function for gene expression, promotor elements. A perspective on higher order DNA structures and their importance on the genetic and genomic levels, as well as tools for their prediction and analysis is presented in the next chapter. Last two chapters deal with the practical application of PCR reactions as a genomic diagnostic tool. Foreclosing chapter discusses the different PCR based approaches and strategies for pathogen identification and offers algorithmic solution based on sequence alignments. The last chapter is a very useful excurse in the theory of oligonucleotide design and a wonderful description of a tool for generation of primers and probes for conventional and more sophisticated PCR and hybridization protocols.

Altogether this book is suitable as a starting guide or a reference gear for basic and more advanced applications in computational molecular biology.

Chapter 1

Mimimum Conflict Phylogeny Estimation in a Nutshell

Georg Fuellen

Contact: Georg Fuellen
Medizinische Fakultät, Arbeitsgruppe Bioinformatik
c/o Division of Bioinformatics, FB Biologie
Schlossplatz 4, 48149 Münster
E-mail: fuellen@alum.mit.edu

1.1 Introduction

The following text, adapted from [1], is a high-level description of the "minimum conflict" phylogeny estimation (MCOPE) method, its motivation and its general idea. Detailed descriptions are given by [2] and [3], and the method is available via [4]; see Box 1 for instructions on how to use the method to generate phylogenetic trees from DNA data.

For a long time, humans have been eager to understand life, as well as themselves, and the roots and origin of both. Modern science places man among all life forms, and tries to describe how these first appeared and have since diversified and changed. Evolution theory describes the initial conditions and the underlying processes. We will assume that this theory has universal validity even though we note that this is impossible to prove. Moreover, many flavors of this theory exist, but we will restrict ourselves to the most basic "undisputed" principles, thereby reducing the probability that we make false assumptions. In fact, we do not need much more than the principle of "descent with modification": Starting with a set (population) of "almost" identical ancestral life forms, we assume that a reproductive mechanism led to generations after generations of descendants, some of these modified with respect to the original. Environmental conditions and other factors separated descendants into subpopulations. After many generations, modifications accumulated such that the reproductive mechanism only works *within* the separated subpopulations, which we may then call "species". If the separation-descent-modification process occurs repeatedly, the result is a phylogeny, that is a hierarchical, tree-like structure that represents the evolution of the various species. Given the species as they are today, our task will be the estimation of the underlying phylogeny; we have to calculate the order of separation events that happened in the past.

Our emphasis will be on the evolutionary processes, that are the processes postulated by evolution theory. The initial conditions (origin-of-life issues) are not investigated in this text, nor are the structures on which they have operated and still operate. These structures, and the possibility to order their substructures sequentially, for example in the form of biomolecular sequences, are taken for granted. Furthermore, we gloss over any variation within a species. Finally, we ignore that some species carry hundreds of blueprints (genes) of some of these structures, since there are mechanisms that keep the blueprints almost identical, like unequal crossing-over and gene conversion. Ignoring variation, we can take one such structure like 18S-rRNA, and talk about "its sequence" and claim that there is exactly one characteristic 18S-rDNA sequence for each species. This sequence is composed of substructures called nucleotides, like any RNA/DNA sequence is. Taking proteins, the substructures would be amino acids. (18S-rRNA is a component of ribosomes, which are part of the apparatus that translates gene information into proteins. It has its own gene, the 18S-rDNA sequence.)

Since we assume that the evolutionary processes are behind all history of life, an investigation of life forms is always an investigation into the results of evolution – "Nothing in biology makes sense except in the light of evolution"[5]. No matter what we are talking about – molecules, genomes, organisms or ecosystems – considering the history of the entities under study can help us a great deal in understanding structure, function and relationships.

In this introduction, we highlight four examples of the productive use of phylogenetic information in biology research – this is "the light of evolution" in action.

- *Molecular level.* The prediction of the three-dimensional structure of proteins can be improved significantly if we know the three-dimensional structure of related proteins. By finding these relatives, phylogeny serves as an aid for molecular modelling, see e.g.[6].

- *Genomic level.* The phylogenetic analysis of viral genes can be exemplified by the case of HIV, the human immunodeficiency virus. The possibility that a dentist has infected his patients has been studied by estimating the phylogeny of the viruses carried by the dentist and his patients [7,8]. The global spread of HIV subtypes is studied phylogenetically in [9]. Cross-species transmission events and the discovery of recombination between viruses are examined in [10]. As a last example, phylogeny inference is used to assess the worldwide variation of HIV and some other viruses in [11]. The phylogeny of many virus families has been studied up to now; the influenza virus is another prominent example, see e.g. [12].

- *Organism level.* The development of body plans can be studied in the light of phylogenetic data. The author of [14] even expresses his hope to establish a "causal relationship" between the evolution of genes called *HOM/Hox clusters*, and the evolution of body plans. The paper already discusses a variety of correlations between gene evolution and organism development.

- *Ecosystem level.* Phylogenies are very useful for the description of biodiversity and the inference of population processes in wildlife, as described in [15]. This paper also discusses how wildlife management, and conservation in general, may benefit from studying phylogenies.

We have highlighted the path of gaining knowledge by considering the evolutionary history of the entities under study. This work is concerned with phylogeny estimation, i.e. gaining knowledge about the evolutionary history itself. The basis for gaining such knowledge is currently expanding at a rapid pace. Due to improvements in nucleotide sequencing technology, larger and larger datasets are in need of phylogenetic analysis, featuring significantly more than just 30 species and just a few hundred nucleotides/amino acids. Instead, for hundreds of species, thousands of nucleotides are now available for analysis. In fact, whole genomes are becoming available, making an all-encompassing phylogenetic analysis possible for the first time. Whole genomes comprise huge datasets on the order of billions of nucleotides, and it would be worthwhile to align the data as far as possible, and to estimate trees from the data that comprise all the inheritable information of the different species.

1.2 Minimum Conflict Phylogeny Estimation in a Nutshell

In the following, we will discuss some general characteristics of our approach.

The first characteristic of our approach is simplicity, and a focus on the most relevant information. We already described our simple model of evolution, which amounts to the hypothesis that separation and descent with modification are the processes that we should focus on. The most natural way of analyzing data resulting from these processes is to look for the modifications. In the case of biomolecular sequences, these are modifications of character states (sequence substructures like nucleotides) that appeared anew in an ancestral species. They testify the exclusive common heritage of all the species to which that ancestral species gave rise. If at least some of these modifications are still visible in the present-day species, we should be able to detect them.

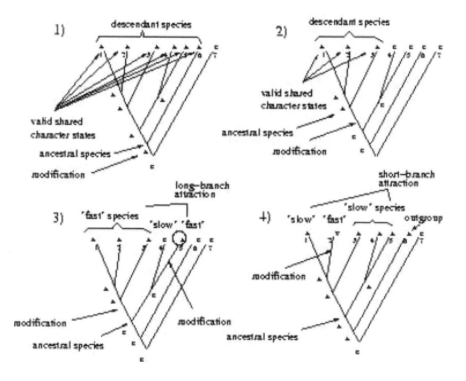

Figure 1-1. Valid shared character states, and branch attraction phenomena.

Consequently, our method for recovering the phylogenetic tree tries to detect character states shared between species because these species are *the sole descendants* of an ancestral species. We argue that these *"valid" shared character states* constitute the most relevant and the least misleading information available. Two examples are given in Figure 1-1, panels 1 and 2, where character states are exemplified by the nucleotide symbols A,C,G,T.

All too often, it happens that character states are shared between some species even though there is no common ancestor from which only these species developed. Instead, these species just feature a different amount of evolutionary change. During their evolution, they were subjected either to much more, or to far less modifications than the others. The former phenomenon leads to "long-branch attraction", and the latter to "erosion", or "short branch attraction". We will describe both phenomena, which are also displayed in Figure 1-1, panels 3 and 4. (Different amounts of evolutionary change are attributed to a "different speed of evolution", and it has become standard to talk about "fast" and "slowly" evolving sequences.) "Long-branch attraction" is the observation of character states shared by "more evolved" sequences because they are modifications that are equal ("convergent") just by coincidence (Figure 1-1, panel 3). "Short branch attraction" is due to character states shared by "less evolved" sequences because they are the leftover of old character states that were modified in the other ("more evolved") sequences only (Figure 1-1, panel 4). Short branch attraction can be visualized by an "erosion" process taking place in the "more evolved" sequences, modifying some of their character states that have been shared before. Since there are usually (but not necessarily) at least a few modifications that coincide with character states elsewhere, at least a low level of long-branch attraction is often a phenomenon that occurs together with short branch attraction. Due to erosion, tree estimation algorithms may be misled by the similarity of the old character states still shared by the uneroded, less evolved sequences.

In general, however, long-branch attraction may happen independently of short-branch attraction, and both may happen independently of the separation-descent-modification process that underlies the phylogeny. For example, the "more evolved" sequences may share similarities due to random convergences, and the "less evolved" sequences may nevertheless have been subject to so many modifications that they feature no shared old character states. Another problematic theme is the possibility that branch attraction happens in parallel with the separation events, that is in concordance with the evolutionary history of the species.

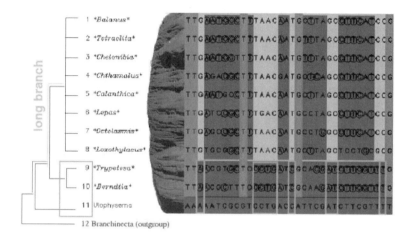

Figure 1-2. Erosion in species 1-8 triggers artifact similarity in species 9-11.

A likely case of short-branch attraction is exemplified in Figure 1-2, which displays an alignment of partial 18S-rDNA from crustacean species on the right, and the putative correct tree on the left. The ancestor of species 1-8 gained many modifications compared to the other species. This erosion in species 1-8 triggers shared old character states in the short-branch species 9-11, in contrast to the valid shared character states that testify the exclusive common heritage of species 1-10. In contrast, the shared old character states do not testify exclusive common heritage of species 9-11 – these species do not have a common ancestor that is not the ancestor of any other species. (It does *not* matter along which branch(es) the larger amount of evolutionary change took place, as long as all species in 1-8 are affected. For example, if they are only recent subjects to significant modifications, we can still observe erosion. In other words, the "caves" in evolved species may consist of conserved modifications, gaps and/or a mix of nucleotides.)

Short-branch attraction can be detected by inspecting a sequence or sequences from species that developed as a side-branch from an ancestor "older" than the ancestral species giving rise to all the descendant species under consideration. This "outgroup" (species 6 in Figure 1-1, panel 4, and species 12 in Figure 1-2) indicates whether character states shared by the short branches only might be old indeed. If they are, we can disregard all artificial evidence that places the short-branch species into one group, and the long-branch species into the other. The decision "old" versus "new" is based on the calculation of "matching rates" with respect to the outgroup. A matching rate compares two sequences (or sets of sequences) and it is calculated by tallying the number of character states that are equal. (For sets of sequences, majority character states are checked for equality.) Basically, the more matching with the outgroup we observe, the more evidence we have for erosion. Matching rates are a very simple concept; after all, simplicity and transparency shall be one important characteristic of our approach. When the analysis is started, the user will need to specify an outgroup in advance. However, we have an unusual freedom in the choice of the outgroup, because we calibrate the matching rate of the shared character states by comparing it with a matching rate tallied over alignment columns that do not feature the shared character states.

To summarize, the second, and possibly the most important characteristic of our method is that it tries to detect short branch attraction – it avoids falling into what we call the erosion trap, and to our knowledge it is the first method that explicitly avoids this systematic error. We even conjecture that as long as the short-branch attraction is stronger than the long-branch attraction artifact, we can detect the former and then avoid *both* problems. This is "Future Work".

The description above is idealistic because usually, there is no clear-cut division between "more evolved" and "less evolved" sequences. Moreover, the speed of evolution may differ in time, across the branches of the tree, resulting in a complex mixture of branch attraction phenomena, and possibly other artifacts. Nevertheless, our method is able to recover correct separations in many cases, as described in [2].

Once we are able to ignore shared old character states, and instead concentrate on the valid ones that indicate *the sole descendants* of an ancestral species, we can do a *heuristic search* of all splits (bipartitions) of the set of species analyzed. Starting with any split, "conflict" may arise if valid shared character states can be found for a subset of the species: if this subset is torn apart by the split, the valid shared character states are then found on both sides. They are torn apart themselves. If a split with no (or minimum) conflict can be found by moving species between the two sides, we assume that we have found the most ancient separation. In hindsight, this most ancient separation divides the set of species into two subsets, and every subset includes *the sole descendants* of one of the two ancestral species into which the species at the starting point (at the so-called root) was separated. Now, our approach can make use of the divide-and-conquer paradigm enabling the fast analysis of large datasets. We have already motivated the necessity of processing speed, and we will now show that divide-and-conquer is a natural ingredient of our approach.

The heuristic search just discussed is designed to reveal the most ancient separation, and the question is how the analysis can be continued. The most natural answer is to use *divide-and-conquer*. We view the two separated sets of species as new problems that can be tackled in the same way. Indeed, in a top-down manner, we explore the hierarchical structure of the dataset by estimating the most ancient separation, followed by the analysis of the two subsets that result from the corresponding split. The two subsets are then analyzed in exactly the same way as the whole dataset was analyzed before; only the outgroup may be different. It is selected in a way that ensures the **most informative** matching rates. Once the subsets are analyzed, we follow up on their minimum-conflict splits, and so on, until the analysis stops with sets of one or two species.

For a completely balanced tree of 100 species, the first divide-and-conquer step divides the problem into two sets of 50 species each. Next, we need to tackle four sets of 25 species, etc. For unbalanced trees, divide-and-conquer slows down, reducing a problem with 100 species to a new one with 99 species, etc. We assume that trees arising from samples of existing species are usually rather balanced. Evidence for this assumption is as follows:

• There is no need to question the overall validity of classic biological systematics, where species are actually classified into many *large* groups of related species on different levels of a classification hierarchy.

• Trees published in the literature are usually quite balanced; few are completely imbalanced so-called "caterpillars".

• Simulation studies usually suggest even more balanced trees, and this "puzzle" is the subject of some recently revived research.

All trees considered are rooted, making both erosion detection and divide-and-conquer possible. These are the third and forth characteristic of our method: We can expect that the algorithm is fast, and we calculate rooted trees (in contrast to less informative unrooted trees) as a side-effect.

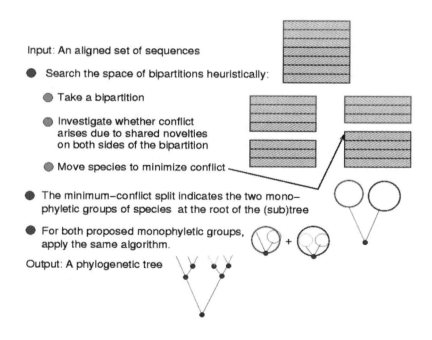

Figure 1-3. Schematic overview of mcope.

A high-level overview of our method, termed "minimum conflict phylogeny estimation" (MCOPE) can be found in Figure 1-3. In general, we aim to model the decision process of a trained systematist who applies a strictly logical approach to phylogeny estimation. Our method may be viewed as followup work that builds upon the logics of phylogeny inference, based on the concepts of *shared novelties* (synapomorphies), *convergences* (homoplasies), and *leftovers* (symplesiomorphies), see e.g.[13]. The relevance of this "cladistic" approach developed by Willi Hennig[16] has been outlined before[17]. We attempt to quantify it, and improve on certain aspects. For example, sigmoid functions are used repeatedly to achieve discriminatory power, e.g. to amplify and filter the evidence found via the comparison of matching rates. The fifth characteristic of our approach is the transparency that comes with its logical foundation: We record the evidence for different hypotheses of phylogenetic relationship, analyse and compare it using simple formulas, and make it possible for the researcher to re-evaluate both the evidence and its analysis.

We believe that it is very important to validate a new method for gaining phylogenetic knowledge. Validation with biological data (in contrast to artificially generated data) is important to prevent circularity, which may occur in subtle ways whenever likeminded researchers write data generation as well as data analysis software. We are tempted to say, "Nothing in phylogeny estimation is validated except in the light of biological data." On the other hand, in the case of artificial data we can be sure to know the correct phylogeny. Therefore, we have done an extensive validation by biological data and artificial data alike. Applied to both kinds of data our method performs very well. In particular, there are several examples where evidence from molecular datasets is now much more in line with morphology-based systematic knowledge [2, 3].

1.3 References

1. Fuellen, G., Computing Phylogenies by Comparing Biosequences Following Principles of Traditional Systematics, Dissertation, Bielefeld 2000.
2. Fuellen, G., Wägele, J.W., Giegerich R., Minimum conflict: a divide-and-conquer approach to phylogeny estimation. *Bioinformatics*. 2001 Dec;17(12):1168-78.
3. Fuellen, G., Wägele, J.W., Giegerich, R. Best Systematist Practice Tranferred to Molecular Data. Organisms, *Diversity and Evolution*. 2001;1:257-272.
4. Fuellen, G., MCOPE Website. http://www.techfak.uni-bielefeld.de/~fuellen/mcope3.html.
5. Ayala F.J., "Nothing in biology makes sense except in the light of evolution": Theodosius Dobzhansky: 1900-1975. *J Hered*. 1977 Jan-Feb;68(1):3-10.
6. Rost, B., Sander, C., Prediction of protein secondary structure at better than 70% accuracy. *J Mol Biol*. 1993 Jul 20;232(2):584-99.
7. Ou, C.Y., Ciesielski, C.A., Myers, G, Bandea, C.I., Luo, C.C., Korber, B.T., Mullins, J.I., Schochetman, G., Berkelman, R.L., Economou, A.N., et al. Molecular epidemiology of HIV transmission in a dental practice. *Science*. 1992 May 22;256(5060):1165-71.
8. Hillis, D.M., Huelsenbeck, J.P., Support for dental HIV transmission. *Nature*. 1994 May 5;369(6475):24-5.
9. Ou, C.Y., Takebe, Y., Weniger, B.G., Luo, C.C., Kalish, M.L., Auwanit, W., Yamazaki, S., Gayle, H.D., Young, N.L., Schochetman, G., Independent introduction of two major HIV-1 genotypes into distinct high-risk populations in Thailand. *Lancet*. 1993 May 8;341(8854):1171-4. Erratum in: Lancet 1993 Jul 24;342(8865):250.
10. Sharp, P.M., Bailes, E., Stevenson, M., Emerman, M., Hahn, B.H., Gene acquisition in HIV and SIV. *Nature*. 1996 Oct 17;383(6601):586-7.
11. Holmes, E.C., Nee, S., Rambaut, A., Garnett, G.P., Harvey, P.H., Revealing the history of infectious disease epidemics through phylogenetic trees. Philos Trans R Soc Lond B Biol Sci. 1995 Jul 29;349(1327):33-40.
12. Scholtissek, C., Ludwig, S., Fitch, W.M., Analysis of influenza A virus nucleoproteins for the assessment of molecular genetic mechanisms leading to new phylogenetic virus lineages. *Arch Virol*. 1993;131(3-4):237-50.
13. Harvey, P.H., Leigh Brown, A.J., Maynard Smith, J., Nee, S. (eds.). New Uses for *New Phylogenies*. Oxford University Press, 1996.
14. Meyer, A. The evolution of body plans; in [13].
15. Moritz, C. Uses of molecular phylogenies for conservation; in [13].
16. Hennig, W. Phylogenetic Systematics. University of Illinois Press, Urbana Chicago London, 1979.
17. Wägele, J.W., First principles of phylogenetic systematics, a basis for numerical methods used for morphological and molecular characters. *Vie Milieu*. 1996;46: 125-138.

Box. 1 Using MCOPE to Generate Phylogenetic Trees from DNA Data.

Minimum conflict can be used via a Web-Form at
http://www.techfak.uni-bielefeld.de/~fuellen/mcope3.html

A version in C is in development, but not finished yet.

The paper in Bioinformatics[2] features a small improvement of the algorithm as described in the dissertation[1] and in the Organisms, Diversity and Evolution paper [3]. The improvement does not have a major effect on trees inferred from the natural data used for testing; most of the time, just the conflict values change a bit numerically. To obtain results as presented in the ODE paper and in the dissertation, in the following please set the parameter *erosion-corrected reliability estimation* to 0 (that is, "OFF"). To obtain results as presented in the Bioinformatics paper, just use the default 1 (that is, "ON"). Other than that, do the following steps to obtain minimum conflict trees:

1. Select the Submission Form

2. Select one of two choices:

   • Select a known dataset (Crustacea, Bilateria, Arthropod, Gnathostomata, Mammalia or Chordata) at "Or select an alignment from the menu:". That way, you can reproduce published results, see the detailed instructions in the text "How to reproduce results from mcope papers" available at the Website.

   • Provide your own dataset, copy-pasting it into the textfield, or loading it via the "Browse" button. Note that very large datasets are not permitted for the prototype Webform; the current limit is 40 species, 4800 columns.

3. Fill in your email address and press the "Submit-Init" button at the bottom of the form.

4. Once the new form is displayed, you need to specify the outgroup unless it is the first sequence. For example, if you provide 10 species, and the last 2 species are the outgroup, type in "9,10" in the outgroup field. Also, specify the erosion-corrected reliability estimation parameter as described above.

5. Press the "Submit" button.

6. The results will take a few minutes to calculate (it's a Perl-Prototype, and graphics calculation as well as a lot of the numerics is done in Perl...). A link to the page of results will appear at the top as well as the bottom of the form. You can inspect the alignment via the "Link to requested parameters". You can watch progress on the "details of submission" page, to which a link is also provided. Eventually, the tree will be printed in newick format, in red color; you can submit it to the Pasteur drawgram interface, if you'd like to have a graphical view. mcope

branchlengths are experimental and undocumented; instead the conflict tables in the "details of submission" page give you hints at the confidence you can have in individual nodes.

Chapter 2

Gene Annotation

Andrea Hansen and Dieter Maier

Contact: Dieter Maier
Biomax Informatics AG, Lochhamer Str. 11, 82152 Martinsried
Germany
Email: Dieter.Maier@biomax.de

2.1 Introduction

Having a raw DNA sequence at hand, one is faced with the task of identifying its distinguishing features and function. In this context, sequence annotation connects raw sequences to information by extracting features resulting from computational prediction, auxiliary biological data and biological knowledge. The DNA sequence contains a multitude of distinguishable features such as regulatory units (see Chapters 8, 9), structural sites or repetitive elements (see Chapter 7); here we focus on genes.

Commonly, the term "gene annotation" is used in a dual sense to mean both the prediction of genes on a sequence as well as the assignment of features and functions to the genes.

In the following text, we describe the tasks faced and the software tools available. The different types of DNA sequences and the corresponding algorithms for gene prediction are presented in Part 1. Part 2 describes the methods and databases used to assign features and functions to a gene. The chapter provides resources to help readers analyze genomic sequences and measure confidence in their results. We do not assume that the reader is familiar with bioinformatics methods and provide only explanations necessary to allow the results to be interpreted. For most of the methods we discuss existing web services where sequences are submitted in an appropriate format and the results are presented directly or via email. Although most methods do not inherently impose size limits on the DNA sequence to be analyzed, one should keep in mind that many services provided via the World Wide Web impose limits for performance reasons. For large and complex analysis projects, setting up a local analysis pipeline may be preferential. However, this will generally require more detailed knowledge about the described methods than is presented here.

2.2 Gene prediction

2.2.1 Gene prediction versus true genes

Although the exact definition of a gene is a matter of debate, one can distinguish certain biologically meaningful concepts, which should be encompassed by any definition. Here we will focus on the molecular definition of a unit of transcription and function, not on the genetically defined unit of heredity and evolution. In this way, a gene will encompass coding and non-translated genes, pseudogenes and anti-sense genes.

From non-translated genes a host of different functional RNAs are produced (for example, tRNA, rRNA, snRNA, miRNA and others). While successful methods for the recognition of tRNA and rRNA genes exist, the development of methods to detect other small RNA genes has just started. Most of these methods are based on sequence similarity between different organisms and subsequent statistical detection of over-conserved genomic regions [1].

For coding genes, the current software does not detect the beginning and end of gene transcription, rather it finds coding regions beginning with a start codon and ending with a stop codon. Thus far few have attempted to extend the range of prediction for eukaryotes to true first and last exon prediction systems. Currently available systems include FirstEF, which predicts 5`-partially coding and untranslated exons [2] and JTEF, which predicts 3`-partially coding but not untranslated exons [3].

For alternative-spliced gene products, the current software usually assembles all possible exons into a colinear gene model. Therefore, detection of alternative splice forms depends on expressed sequence tag (EST) and cDNA data.

2.2.2 Types of sequences

The sequence type determines the gene-prediction method that should be used for coding genes. The main distinctions here are among intronless genes from mostly coding prokaryotic and lower-eukaryotic genomic DNA and higher eukaryotes with intron-rich genes and mostly non-coding genomes. EST sequences are a special case; the challenge of finding the correct open reading frame (ORF) stems not from multiple possibilities but from low sequence quality, which may create artificial stops and frame shifts. Most algorithms focus on one of the above-mentioned types of DNA. In addition, algorithms need to be trained (see below) to recognize the organism-specific features of a gene. Currently, trained algorithms for vertebrates, invertebrates, plant, fungi, gram-positive bacteria and gram-negative bacteria are available from Web servers. If the DNA to be analyzed has a very unusual composition, it may be necessary to create a special training set, otherwise the gene prediction results will be much less reliable.

For non-coding genes, there are similar differences between prokaryotes and eukaryotes. However, most of the existing algorithms allow the analysis of both types of sequences using different internal parameters.

2.2.3 Non-coding gene prediction

As mentioned above two types of non-coding genes, tRNA and rRNA genes, can be readily detected. For tRNA genes, the algorithm tRNAscan-SE [4] provides a sensitivity of 99.5% (missing less than one out of 100 existing tRNA genes) and a selectivity of practically 100% (less than one false positive in 15 gigabases (Gb), the equivalent of three human genomes).

For rRNA genes, the high degree of conservation among organisms and the size of these genes allow detection by sequence similarity searches of collections of known rRNA gene sequences, such as the European rRNA database [5].

For other non-coding RNA genes, which are mostly transcribed into regulatory RNAs such as spliceosomal RNAs, microRNAs or small nucleolar RNAs, a variety of detection approaches exist. One approach is to search for initiation, termination and processing sites (e.g., splice sites) of RNA transcription and select all resulting putative RNAs without an ORF. However, as detailed below, the ab initio detection of transcriptional units is, thus far, an unsolved computational problem. The difficulty is further enhanced by the low compositional bias (see below) of non-coding genes, which provide one of the strongest signals for the detection of coding genes. Therefore, the most powerful current approach uses genomic sequence similarity searches. With this approach, a statistical test can detect function-conserving compensatory base changes, which distinguish non-coding genes from conserved intergenic regions [1].

2.3 Coding gene prediction

2.3.1 Methods

Historically two approaches to gene prediction exist. One approach is based on the detection of sequence similarities between genomic (or translated genomic) sequences and EST/cDNA or protein sequences (called extrinsic). The other approach focuses on ab initio prediction (called intrinsic). While sequence similarity based gene prediction exhibits a high degree of accuracy, it depends on the availability of sufficiently conserved sequences, equivalent to BLAST scores of e-120 and below [6]. Without support from other methods, useful sequences are restricted to the same or very closely related organisms.

Ab initio prediction, on the other hand, suffers from its reliance on statistical and, therefore, inherently fuzzy features. In recent years, the merger of both approaches has greatly increased the breadth and quality of gene prediction as is detailed below.

2.3.2 Ab initio prediction

Almost all current gene prediction algorithms for intron-rich genomic DNA use Hidden Markov Models (HMMs) to build a probabilistic model of a gene. The same methods are used for prokaryotic DNA, but there are also successful algorithms that rely on detection of ribosomal binding sites (RBS) and skewed-base distribution, detected by weight matrices and hexamer frequencies.

One of the reasons why HMMs are so successful for more complex eukaryote sequences is the ease with which they are able to integrate probabilities for certain states computed by separate modules. A gene prediction HMM might integrate information from specific sub-HMMs for exons or introns, neural networks for splice site prediction, hexamer frequencies, BLAST alignment scores, etc.

HMMs were originally developed in speech recognition. For a detailed description regarding HMMs and their use in gene annotation see for example [7]. Briefly, an HMM explicitly assigns probabilities to states and transitions of a graph. In gene prediction, the states refer to sequence features such as untranslated exons, start exons, internal exons, introns, splice sites and untranslated regions (UTRs). Transitions between states then refer to whether a base belongs to the same feature as the previous base or a new state has been reached (see Figure 2-1). The order of the Markov model determines how the environment influences the probability for a base to belong to a certain state. That is, in a model of order 0 the probability of nucleotide v+1 only depends on nucleotide v, while in a 5th order HMM, the previous 5 bases influence the probability. As mentioned above, HMMs are especially well-suited to integrate results from independent algorithms that calculate the probability of a certain state as long as the output of the algorithm can be stated as a probability.

After setting up the principal model of all existing states and allowed transitions (e.g., there is no transition from intron to 5' UTR) the HMM must be trained to adjust its probabilities.

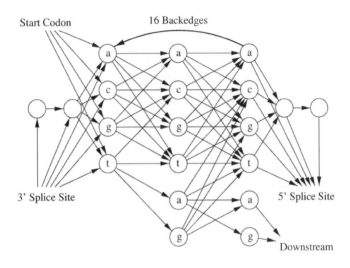

Figure 2-1. *Example of an HMM exon model taken from Henderson et al.* [66]. *Entry points into the exon model are start codon and 3'Splice site, which can be delivered from separate models. Within the exon nucleotides are emitted as codon and loop back after each triplets. Exit points are the models for 5'Downstream state from the stop codon and 5'Splice site. As Splice sites can occur within codon triplets, two general states at begin and end of the exon allow entry into the codon triplets from every reading frame.*

The training data needs to provide a balance between coherence, which allows the extraction of determining properties such as weight matrices, and diversity, which ensures that a maximum of possible values is represented and no overfitting occurs. The gene models from the training set should be supported by experimental evidence as much as possible. The probabilities of the HMM will be adjusted using a machine-learning approach such as the Expectation-Maximization (EM) algorithm. The training set and resulting probability matrices determine the kinds of sequences that can be submitted for gene prediction (e.g., if the query sequence is from a plant, a human training set will give inferior results).

Analyzing a new sequence requires aligning each base in the sequence to a particular state of the model (e.g., exons or introns). Viewed in another way, the HMM creates a path of states from the raw sequence. To select the most likely path from the various possibilities, most methods use the dynamic programming Viterbi algorithm.

At least two sub-models are commonly used with such an HMM: one for exon and intron recognition, the other for splice-site detection. Within exons, regularities of codon usage and codon-codon correlation give rise to frame-specific hexamer frequencies with a biased distribution between coding and non-coding DNA. In addition, preferential codon usage creates a periodicity for same bases at every third position within a coding region. In prokaryotes these features are in themselves enough to distinguish long intron-less ORFs from inter-genome regions.

For splice site recognition, simple models use a positional weight matrix for donor- or acceptor-site recognition. However, a weight matrix does not account for the correlation between positions of a signal. To achieve this, more elaborate models have been created including weight arrays and Markov chains to represent adjacent correlations, decision-trees or maximal-dependence decomposition for non-adjacent correlations, and neuronal networks for arbitrary, nonlinear dependencies. For recent reviews on gene prediction methods see Zhang [8] and Mathe et al. [9].

2.3.3 Use of sequence similarity

There are several ways, and combinations thereof, in which sequence similarity is commonly used within the gene-prediction paradigm.

• Models can be determined solely according to their similarity with other sequences.

• Ab initio gene predictions can be refined using evidence from sequence similarity.

• Parameters for ab initio prediction can be calculated from similarities between prediction and similar sequence.

Depending on the envisioned usage, a number of different sequences are scanned for similarity:

• The genomic sequence (or its six-frame translation) compared to databases of ESTs/cDNA (or six-frame translations)

• A six-frame translation of the genomic sequence compared to a database of protein sequences

• The protein translation from predicted gene models compared to a database of protein sequences

• The genomic sequences compared to a genomic sequence from a closely related organism

2.4 Quality of prediction

For prokaryotic genomes the problem of gene prediction has essentially been solved, with prediction accuracy of 95 to 99% on the gene level [11, 12, 13]. Here, accuracy describes the combined measure of sensitivity (the percentage of true genes which have been correctly predicted, i.e., all exons with correct start-, stop-, and splice sites) and specificity (the percentage of predicted genes which are true). Even so, several minor problems remain, e.g., distinguishing artificial and true overlapping genes, gene number over-prediction in GC-rich genomes (due to a statistical underrepresentation of AG rich Shine-Delgano sequences indicating a start codon and AT rich stop codons ORFs overlap strongly) or reliable detection of peptides with less than 80 amino acids.

However, for eukaryotic genes with large introns in gene-scarce genomes, accuracy is much lower. Using test sets with single genes per contig as a benchmark, the best ab initio algorithms have a gene-level accuracy of approximately 40%, while they reach about 75% at the exon level and 90% at the nucleotide levels [13]. Using multi-gene contigs however, these numbers drop to about 5% accuracy at the gene level. Combining several gene prediction algorithms improves accuracy by about 10% [6, 15]. Integrating external evidence, such as protein similarity and EST/cDNA alignment, to the ab initio prediction also increases accuracy by about 10% on the exon and gene levels. Recently, the growing number of available genome sequences has allowed a new approach in which evidence from genome comparison is integrated with ab initio gene prediction. This approach also improved prediction accuracy for multi-gene sets by about 10% on the gene level [16, 17].

Using a statistical combiner algorithm, Allen et al [17] integrated results from splice-site predictions and gene prediction, from ab initio and genome alignment enhanced algorithms, with protein and EST/cDNA alignments. In this way, they reached an impressive accuracy of about 75% on a test set of 1783 *Arabidopsis thaliana* genes (distributed in roughly 80 megabases (Mb) of genome sequence). However, note that the *A. thaliana* genome has a higher gene density and more compact genes than (for example) *H. sapiens* and gene prediction is less challenging. This is reflected in the accuracy statistics of the genome-alignment-enhanced gene-prediction tool, Twinscan [15], which achieved 67% accuracy using the test set and 15% accuracy using 17,000 human RefSeq sequences (a set of cDNA supported gene models) within the 3 Gb human genome as benchmark. Although eukaryote gene-prediction accuracy has improved in recent years, the current approaches often result in partial genes, fragmented genes, gene fusions and spurious predictions (see Table 2-1).

Method	Problem
Ab initio	Poor sensitivity and specificity on gene level, whole genes or exons are missed or wrongly predicted
Similarity to existing expressed sequence tags (ESTs)	Contaminating ESTs derived from unspliced mRNA, genomic DNA and nongenic transcription
Similarity to existing gene/proteins	Difficult to distinguish pseudogenes (non-protein coding) and novel genes
Similarity to other genome sequence	Difficult to distinguish pseudogenes and regulatory regions

Table 2-1. *Quality overview.*

Keeping this in mind, one can be reasonably confident of results on the exon level, but not overestimate the accuracy on the gene level.

2.5 Interpretation of results

Algorithms present their results in different ways. While some provide only a list of coordinates on the input DNA sequence, most extract a translated protein and coding DNA for their predictions, usually in FASTA format. A generic gene-finding format (GFF) has been proposed to allow simple comparison of results from various algorithms (http://www.sanger.ac.uk/Software/ formats/GFF/GFF_Spec.shtml). The format provides the source of the feature, the feature description and its coordinates. Coding sequence and translation are not mandatory, but can be included as a comment.

Another common output includes graphical visualization of exons and gene models mapped to the genomic sequence. Some algorithms add evidence scores to the output, providing a measure of confidence for each exon. For quick comparison of the output from different algorithms, one can create sequence alignments from the predictions or use a visualization tool. Open source viewers such as Apollo [18] allow one to overlay several gene models and sequence alignments and to edit the gene models manually. In light of recent automatic evidence-integration methods, manual curation may seem outdated. However, the well-established inaccuracy of pure computational methods indicates that it is still worth the effort. Various genome annotation projects show that manual curation improves automatic predictions, even from integrated methods, and enhances accuracy by up to 20% on the gene level (our unpublished results, [19].

2.6 Coding gene prediction algorithms

2.6.1 Prokaryotes and lower eukaryotes

Three well performing algorithms are described below. They can be accessed via the Web or installed locally.

GeneMark.hmm [11] is based on an HMM which simultaneously generates ORF predictions on both strands, prohibiting overlapping ORF predictions. A post-processing step searches for ribosomal binding sites (RBS) and extends ORFs, thus allowing for overlaps. It is based on and extends the GeneMark Markov models for coding and non-coding sequences. A version trained for sequences from higher eukaryotes is available too.

GLIMMER [12] uses an Interpolated Markov Model (IMM) which enables the Markov model to adapt its order (the size of the environment sampled to estimate a probability) according to the context encountered.

ORPHEUS [10] is a combinatorial tool integrating statistical and alignment information. It first performs a similarity search of the translated genome sequence against a protein database. Subsequently, ORFs are extracted starting with global protein alignments as seeds to define codon-usage statistics and create a RBS weight matrix for the given organism. Regions with high coding potential are used as seeds for ab initio ORF prediction. ORF length is optimized for coding potential and RBS strength. ORF overlaps are resolved according to length and coding potential.

2.6.2 Eukaryote — ab initio algorithms

Genscan[20], **GeneMark.hmm**[11] **Genie**[21] and a host of other HMM-based gene-prediction software has been developed during the 1990s and are available as Web servers. As described above, the performance of the algorithms mentioned is approximately equivalent and accuracy reaches up to 40% on the gene level for single gene contigs. However, for multi-gene contigs accuracy falls to about 5% on the gene level. Therefore, algorithms that integrate information from alignments are better suited to the problem. Unfortunately, so far, combinatorial servers, which integrate results from several gene prediction algorithms and enhance accuracy, are scarce; **GeneComber**[14] and **EGPRED** (http://www2.imtech.res.in/ raghava/egpred/) are notable exceptions.

2.6.3 Eukaryote — alignment enhanced algorithms

A number of algorithms exist which provide some integration of ab initio prediction and sequence-similarity evidence. As described above, algorithms integrating ab initio models with EST, cDNA or protein sequence similarity are most useful in a medium range of sequence similarity; that is, similarities with BLAST expectation values (E values) below e-20 will show no notable improvement of the model, while similarities above e-120 can be used without ab initio prediction. For genome sequences, again useful similarity is limited to overall identity below 80% , which is equivalent to the evolutionary distance between human and mouse or Arabidopsis and other dicotyl plants. Most existing algorithms are not adapted to higher similarities and will produce spurious exon predictions from non-coding conserved regions. Evidence from sequence similarity can be used in several ways to refine existing models, to refine prediction parameters or to assemble gene models according to the best alignment path.

2.6.4 Refining ab initio predictions or parameters using similarity evidence

FGENESH++C integrates the HMM-based ab initio gene prediction from **FGENESH**[22], with sequence alignments against RefSeq gene models, protein and EST/cDNA sequences. Ab initio predictions are automatically adapted and predicted exons having sequence similarity receive additional weight. Unfortunately this combinatorial method is not available as a Web server, instead only the separate methods **FGENES+** with protein-alignment integration and **FGENESH-C** with EST/cDNA-alignment integration can be accessed. Another version of the software, **FGENESH-2** allows the integration of comparative genome alignments.

GenomeScan[23] integrates a Genscan-based ab initio exon–intron and splice-site model with similarity to protein sequences. It optimizes the resulting gene prediction for maximum probability based on available similarity information. The GenomeScan Web server offers Genscan versions trained for vertebrate, Arabidopsis or Maize sequences. Users need to provide appropriate protein sequences for alignments. To this end, one can do a *blastx* search with the genomic sequence or use the peptides predicted by Genscan for a *blastp* search.

Twinscan [15] is based on a reimplementation of Genscan, which is extended by the integration of cross-species similarity into its probability model. Various probabilities are assumed for conservation in coding regions, translation initiation and termination signals, untranslated regions, and introns or intergenic regions. The Web server offers versions trained for human, non-human mammals, Arabidopsis, other dicotyl plants, *Caenorhabditis elegans*, other Caenorhabditis, and *Cryptococcus neoformans*. *blastn* is used to align the query sequence to the determined target sequence.

2.6.5 Algorithms that simultaneously align sequences and predict genes

SGP1 (Syntenic Gene Prediction) applies a two-step procedure of sequence alignment and exon-candidate prediction. Exons are rescored and assembled into genes according to the sequence alignment. As SGP-1 emphasizes similarity, it is best suited for conserved genes and species with an evolutionary distance such as that between human and chicken [24]. Exon candidates are generated by scanning for start- and stop codons, and splice sites using a likelihood-profile pattern. Currently, profiles for vertebrates and angiosperms are available.

Doublescan [16] uses a paired HMM to simultaneously predict genes for and align two syntenic sequences. The sequences need not be prealigned by the user for submission to the Web server; however, it should be ascertained that the sequences are collinear (e.g., by performing a *blastn* search). The Web server currently offers a version trained for human and mouse sequences. Sequences are analyzed as submitted; no automatic analysis of the reverse-complemented sequences is provided.

SLAM [25], like Doublescan, uses a generalized paired HMM to simultaneously predict genes for and align two syntenic sequences. In addition, the model explicitly distinguishes coding from non-coding sequence conservation. The Web server expects collinear sequence pairs as input and is configured for human - mouse sequence input. The server page states that sequences from other organisms with similar evolutionary distance can also be submitted. However, because probabilities for initiation, duration and termination of coding and non-coding states as well as for splice site prediction have been derived from human sequences, sequences from non-vertebrates will probably require a new parameter set.

2.6.6 Combinatorial servers

EGPRED integrates ab initio predictions from GenScan and HMMgene with splice-site predictions from GeneSplicer [26] and sequence similarity searches compared to protein and intron sequences.

Combiner [17] uses a statistical combinatorial algorithm and has been described above. It is currently not provided as a Web server but is available for local installation. It can integrate all evidence that is provided as list of coordinates, and is therefore very versatile with respect to the number and nature of possible methods for integration.

2.6.7 Gene prediction for EST/cDNA

ESTScan [27] uses a 5th order HMM for coding-region detection, the same model used in the Genscan algorithm. It recognizes the bias in hexanucleotide usage found in coding regions relative to non-coding regions. The HMM has been extended to allow for various sequencing errors such as ambiguities, insertions and deletions, and corrects frame shifts (by inserting a "base" X) when such correction improves the coding region statistics.

2.6.8 Similarity-based gene prediction

GeneSeqer [28] creates spliced alignments of EST/cDNA sequences against genomic DNA by dynamic programming. The Web server at GeneSeqer@PlantGDB [29] provides a large plant EST/cDNA database and integrates it with splice-site prediction and sequence alignments to generate similarity-based gene prediction in plants.

Sim4 [30] aligns EST/cDNA sequences against genomic sequences, allowing for splicing and sequencing errors. Originally intended for sequences from the same organism, it has been shown to provide significant results across evolutionary distances as large as human to mouse. Starting from short ungapped alignments (high-scoring pairs), sim4 creates a path of alignments and searches for splice sites using a simple GT...AG or CT...AC pattern.

GeneWise [31] combines a similarity search of protein sequences compared to translated genomic sequences with an error-correcting functionality that allows sequencing errors and frame shifts, and uses splice-site detection to fine-tune the alignment. It extracts an exon–intron model from all six frames of a genomic sequence that is restricted to the aligned region and does not necessarily cover start or stop codons. As mentioned above, for proteins from the same or a closely related species, this creates a highly reliable gene model.

2.7 Gene feature prediction

After identifying the protein-coding part of the genomic sequence, the next step is to provide functional prediction. There are several ways to provide functional prediction, which will be listed and explained in the following section. Most of the tools require a gene prediction before they can be used to annotate a sequence.

There is no strict order to annotate a gene. The described methods can be used in the given order or rearranged depending on the annotator's experience, the sequence and the time available.

Most of the annotation methods are based on a similarity search. The sequence is compared to various databases to find similar sequences, and the functions of the match are assigned to the sequence. That is, a function can be assigned to the sequence only if a sufficiently similar sequence in the database has already been annotated.

Gene annotation tools are based on the assumption that the biological properties of a particular protein are shared within a family of proteins and are characteristic for the family based on a common ancestor. Without the assumption of shared sequences, patterns, structures, etc., functions cannot be assigned to an unknown protein with in silico methods.

2.8 Homology search

The simplest way to get information about a gene is to do a similarity search of a sequence database using local alignment methods. The most common similarity-search tools are BLAST [32,33] and FASTA [34,35]. Both tools use heuristic approaches for fast database searches, and generate a local alignment with the query sequence and the match sequence in the database. Here, we will focus on BLAST, but similar rules apply to FASTA. For a comparison of the two methods, see [35].

Before performing a database search, you should consider which BLAST method to use and which database to search.

2.8.1 BLAST methods

There are three basic BLAST methods: *blastn*, *blastx* and *blastp*. *blastn* compares a nucleotide sequence to a nucleotide database. *blastx* translates a nucleotide sequence in all six reading frames and compares it to a protein database. This method is useful if the correct reading frame or the orientation of the sequence is not known. *blastp* compares a protein sequence to a protein database. *blastp* is the method of choice if you are relatively certain that the gene prediction algorithm has given a clear result and one distinct reading frame translated in the correct protein sequence. A prerequisite to this method is a sequenced clone with cDNA-like quality.

Why are different kinds of BLAST searches necessary? Various kinds of questions are asked. The main question may be: *What is the function of my unknown protein?* Other questions may follow: *Which is the correct reading frame in the nucleotide sequence? Where is the transcription start of the protein? Is there more than one exon in the sequence? Are there known paralogous or orthologous genes?* For some questions the nucleotide sequence must be used when performing the database search, for others, the protein sequence.

2.8.2 Databases

Based on the BLAST method, an appropriate database must be chosen. The question is not whether to choose a nucleotide or protein database; the kind of database is already determined by the BLAST method. The question is which nucleotide database or which protein database. The selection of the database influences the results. For example, running *blastn* against an EST or cDNA database will give information about the transcription start in the sequence. Running a similarity search of a curated protein database such as SWISS-PROT [36] or UniProt [37] may yield more reliable hints to the protein function than running a search of GenBank [38]. On the other hand, a match in a particular database can only be found if the database contains an appropriate sequence with similarity to the query. If a database search of SWISS-PROT gives no matches, you may have to perform a search of all genes or proteins. Most database searches start by querying the whole database.

Currently, nearly any database can be used to perform a similarity search on the Web. The most common website for performing a BLAST search is the National Center for Biotechnology Information (NCBI) website. Besides performing a BLAST search against all sequences in the GenBank one can select several specialized databases, such as the EST database, dbEST [39], and genome-specific (e.g., human or mouse) and organelle-specific databases.

2.8.3 Interpreting the results

Unfortunately, there is no general rule for interpreting the significance of BLAST results (i.e., there is no significance threshold for the E value). Loose thresholds can be defined for raw filtering of the matches, but the final filtering must be done manually.

2.9 Pattern search

What happens if the sequence comparison returns no results? Either the protein is not in the database or the similarity to other proteins is too weak for the method. In the latter case, a pattern search should be performed. Query pattern databases with the sequence to find the protein family to which the sequence belongs. Pattern databases group proteins in protein families based on common patterns. A pattern is a common attribute that significantly characterizes a protein family; the pattern can be very simple, as a regular expression (also called a signature), or more complicated, as in a profile or a Hidden Markov Model. For all kinds of patterns, a multiple alignment of related proteins must be performed. Usually, the multiple alignment is manually edited and called a seed alignment. Using the multiple alignment, the pattern is calculated with various algorithms.

The following section describes the kinds of patterns characteristic for particular databases.

2.9.1 PROSITE patterns

Currently, the PROSITE database [40] contains 1,235 entries that describe 1,676 different patterns and profiles (Release 18.18). The starting point for the patterns is the multiple alignment of related proteins. Instead of returning a consensus sequence of the multiple alignment, which gives the most common letter for any position in the sequence, a pattern similar to a regular expression, which describes all allowed letters in any position in all the sequences, is returned (see figure 2-2). Such patterns are not calculated for the whole sequence, but for a short well-conserved region of the protein. Typical regions for such patterns are distinct functional regions in a protein (e.g., the catalytic site).

Seq 1	R V T I G H A Q R G
Seq 2	R K N N G H M Q Q G
Seq 3	R T C P G H V Q R G
Seq 4	R A V T G H T Q R G
Consensus	R X X X G H X Q R G
Pattern	R-(X)$_3$-G-H-X-Q-[QR]-G

Figure 2-2. A multiple alignment of four sequences with the consensus sequence and the corresponding signature for the functional region of the proteins: Position 1 must be R, Positions 2 to 3 can be any amino acid, Position 5 must be G, Position 6 must be H, Position 7 can be any amino acid, Position 8 is Q or R and Position 10 must be G. In this way, all allowed amino acids for any position in the pattern are described, providing more information than a consensus sequence.

33

Why is such a database useful for annotating a sequence? All entries are linked to annotation documents containing information on the protein family that belongs to the pattern. The database offers a tool for scanning all patterns with the sequence, ScanProsite. Using ScanProsite, a protein sequence can be scanned for the occurrence of patterns and profiles stored in the PROSITE database, and protein databases can be searched with a pattern provided by the user [41]. The program PRATT [42,43] can be used to generate patterns.

2.9.2 PROSITE profiles

In addition to signature-based searches, PROSITE can perform profile-based searches. A profile contains multiple-alignment information as scores for each amino acid at each position (calculated by a substitution matrix) and gap costs for deletions and insertions. That is, for each position of the multiple alignment, 20 scores are given — one for each amino acid; the more an amino acid is conserved in the multiple alignment, the higher the score for the amino acid. The profile provides a rating for each amino acid at each position.

As previously described for a pattern search, ScanProsite can be used to search the profiles stored in the PROSITE database.

2.9.3 PRINTS and BLOCKS motifs

The PRINTS [44] and BLOCKS databases [45,46] consist of fingerprints and blocks, respectively. These patterns are multiply aligned, ungapped segments corresponding to the most highly conserved regions of proteins. Each entry consists of several fingerprints or blocks, which are characteristic for a protein family. Both databases provide tools for searching protein families with a sequence: FingerPRINTScan in the PRINTS database [47] and Block Searcher in the BLOCKS database.

2.9.4 Pfam HMM-profiles

Hidden Markov Model profiles present a scoring scheme similar to that of profiles but based on probabilities instead of scores. Such patterns are the basis of the Pfam (Protein families database of alignments and HMM) database. The Pfam database is a comprehensive collection of multiple sequence alignments and Hidden Markov Model profiles covering common protein domains and families [48]. The HMM profiles are constructed in two steps: a multiple alignment is calculated with a manually curated set of sequences representing a protein family; this seed alignment is then used to create a global and a local HMM profile.

The HMMER software performs the calculation of the HMMs and searches using a sequence in the PFAM database [49].

2.9.5 Meta database InterPro

Instead of scanning all pattern databases separately, a more comprehensive search of the InterPro database can be performed. InterPro aims to provide an integrated view of all protein-family databases; a comprehensive set of annotations has been created by merging the information from each member database [50].
InterPro provides a service called InterProScan [51], which can be used to search all pattern databases simultaneously. It combines various pattern search methods in one tool. The current release of InterPro (release 7.1) covers the following databases: SWISS-PROT, PRINTS, TrEMBL, Pfam, PROSITE, ProDom, Smart, TIGRFAMs and PIR.

2.10 Functional annotation

Functional assignments can be given to sequences using existing databases or catalogs with distinct terms or categories.

2.10.1 Enzyme nomenclature (Enzyme Commission (EC) numbers)

The Nomenclature Committee of the International Union of Biochemistry and Molecular Biology (NC-IUBMB) in consultation with the IUPAC-IUBMB Joint Commission on Biochemical Nomenclature (JCBN) has classified all enzymes in six main categories. These categories can be used to classify every enzyme. There are several databases available in the Internet for browsing the EC numbers:

- The **IUBMB Enzyme Nomenclature** is the oldest database with the EC catalog.

- **ENZYME** is a repository of information related to the nomenclature of enzymes. It is primarily based on the recommendations of the NC-IUBMB and describes each type of characterized enzyme for which an EC number is provided [52].

- The **Kyoto Encyclopedia of Genes and Genomes (KEGG)** describes enzymes with corresponding EC numbers in the context of the biochemical pathway in which the enzymes are involved.

A catalog with EC numbers is useful for assigning a function to an enzyme, but not all proteins are enzymes. For this reason, it was necessary to develop new classification catalogs, which can be used to assign functions to other proteins. The most popular classification catalogs are FunCat and GO.

2.10.2 Funcational Catalog (FunCat)

The FunCat is an annotation scheme for the functional description of proteins from prokaryotes, unicellular eukaryotes, plants and animals. Taking into account the broad and highly diverse spectrum of known protein functions, the FunCat consists of 30 main functional categories (or branches) that cover general fields such as

35

cellular transport, metabolism and signal transduction. The main branches exhibit a hierarchical, tree like structure with up to six levels of increasing specificity. FunCat was developed by the Munich Information Center for Protein Sequences (MIPS) [53].

2.10.3 Gene ontology (GO)

The Gene ontology (GO) project provides structured, controlled vocabularies and classifications that cover several domains of molecular and cellular biology and are freely available for community use in the annotation of genes, gene products and sequences. The project began as a collaboration between three model organism databases: FlyBase (Drosophila), the Saccharomyces Genome Database (SGD) and the Mouse Genome Database (MGD) in 1998. Since then, the GO Consortium has grown to include many databases, including several major repositories for plant, animal and microbial genomes [54].

2.11 Localization prediction

The prediction of a protein's cellular and subcellular localization can be helpful for functional annotation. Several automated localization predictors have been developed and made available online.

Given the amino acid sequence information alone, **PSORT** uses various kinds of information organized as "if-then" rules for predicting protein localization sites in Gram-negative bacteria. It considers four localization sites: the cytoplasm, the inner (cytoplasmic) membrane, the periplasm and the outer membrane [55, 56]. **PSORT II** is optimized for eukaryotes. Proteins that are sorted through the so-called vesicular pathway usually have a signal sequence in the N-terminus, which is either cleaved off after translocation through the endoplasmic reticulum (ER) membrane or not and remains as a transmembrane segment [57].

iPSORT is a subcellular localization site predictor for N-terminal sorting signals. Given a protein sequence, iPSORT will predict whether the sequence contains a signal peptide, a mitochondrial targeting peptide or a chloroplast transit peptide. The structure of iPSORT is a decision list, which works with two rules: an amino acid index rule (based on the biochemical properties of the particular amino acid) and an alphabet indexing/pattern rule. The latter can be considered to be a discrete, non-ordered version of an amino acid index. The original amino acid sequence is converted to a string of 0s, 1s, and 2s. To be judged "yes" to a given node, the converted sequence must match a certain pattern, within a certain substring [58].

TargetP is a neural-network-based tool and was developed for large-scale subcellular location prediction of newly identified proteins. Using N-terminal sequence information only, it discriminates between proteins destined for the mitochondrion, the chloroplast or the secretory pathway and their localizations [59].

SignalP predicts the presence and location of signal peptide cleavage sites in amino acid sequences from different organisms: Gram-positive prokaryotes, Gram-negative prokaryotes and eukaryotes. The method incorporates a prediction of cleavage sites and a signal peptide/non-signal peptide prediction based on a combination of several artificial neural networks and Hidden Markov Models [60].

In addition to TargetP and SignalP, the CBS Prediction Server offers several tools for the prediction of localization and cleavage sites.

2.12 Protein structure prediction

Once a protein sequence has been determined, the more difficult task is to make a structure prediction. The structure prediction helps to elucidate the protein function (e.g., in drug design). Because experimental methods for determining protein structure are very time-consuming and expensive, computational prediction methods are needed.

Protein structure prediction can be done using one of several techniques. Prediction by **comparative modeling** exploits the tendency that evolutionarily related proteins with similar sequences have similar structures. For such prediction methods, sequences with a known structure that are very similar to the sequence in question are needed. After a multiple alignment of these sequences is constructed, it is used to obtain the structural model by copying the backbone elements of the template and adding loops and side chains. To obtain similar sequences with a known structure, a BLAST search of a structure database, e.g., the Brookhaven Protein Data Bank (PDB) can be performed. Another tool is the SWISS-MODEL, a fully automated protein structure homology-modeling server, which is accessible via the ExPASy Web server [61].

Threading methods compare a target sequence to a library of structural templates to produce a list of scores. The scores are then ranked; the fold with the best score is assumed to be the one adopted by the sequence. Structure prediction by threading can be performed if similar sequences with known structure are not available. Threading Metaservers like the Fold prediction Metaserver provide simultaneous prediction by several methods.

Finally, **ab initio** prediction tries to model the energetics involved in the process of folding and then find the structure with lowest free energy. A tool for an ab initio prediction is PETRA [62].

2.13 Automatic prediction server

To use more than one method for prediction of a sequence, it is recommended to use one of the prediction servers on the Web.

The **PredictProtein** Server is a good starting point for predicting the structure of a protein. PredictProtein is an automatic service for protein database searches and the prediction of certain aspects of protein structures [63].

Panal is an integrated resource for protein sequence analysis. The tool allows the user to simultaneously search a protein query sequence for motifs from several databases, and to view an intuitive graphical summary [64]. The results of Panal can be visualized with **Metafam** which gives an indication of the confidence of each assignment.

Pedant provides exhaustive automatic analysis of genomic sequences using a wide variety of established bioinformatics tools through a comprehensive Web-based user interface [65].

Pix is a Web tool for viewing results when many peptide analysis programs are used to analyze a peptide sequence.

Acknowledgements

We would like to thank our colleagues Kaj Albermann and Sheridon Sanford for discussion and improvements on the text.

Appendix 2-1 Web links

necomber/ GeneComber combinatorial

Prediction Server
http://www.sanger.ac.uk/Software/analysis/doublescan/ Doublescan Server (Human-mouse only)
http://soft.ice.mpg.de/sgp-1/ SGP-1 Server (0,15 Mbp, 1,0 Mbp as e-mail server)
http://genes.cs.wustl.edu/ TwinScan (Human, mammal, A. thaliana, dicotyl plant, C. elegans)
http://www2.imtech.res.in/raghava/egpred/ EGPRED Server

EST/cDNA + Alignment
http://www.ch.embnet.org/software/ESTScan.html ESTScan Server
http://www.isrec.isb-sib.ch/ftp-server/ESTScan/ ESTScan download
http://bioinformatics.iastate.edu/cgi-bin/gs.cgi general GeneSeqer Server
http://www.plantgdb.org/cgi-bin/PlantGDB/GeneSeqer/PlantGDBgs.cgi Plant-specific GeneSeqer Server
http://pbil.univ-lyon1.fr/sim4.php Sim4 single sequence mode Server
http://sky.bsd.uchicago.edu/batch_sim4.htm Sim4 batch mode Server
http://www.ebi.ac.uk/Wise2/ GeneWise Server

Other
http://igs-server.cnrs-mrs.fr/igs/banbury/ Benchmarking Server 1997
http://www.cs.ubc.ca/~rogic/evaluation/results.html Rogic Benchmarking 2002
http://www.nslij-genetics.org/gene/ Bibliography on Computational gene recognition
http://www.tigr.org/software/combiner/ Combiner download

Homology search
http://www.ncbi.nlm.nih.gov/BLAST/ BLAST
http://fasta.bioch.virginia.edu/ FASTA
http://www.ebi.ac.uk/swissprot/index.html SWISS-PROT
http://www.ebi.uniprot.org/UniProt/index.shtml UniProt
http://www.ncbi.nlm.nih.gov/ GenBank

Pattern search
http://us.expasy.org/prosite/ PROSITE
http://ca.expasy.org/tools/scanprosite/ ScanProsite
http://ca.expasy.org/tools/pratt/ PRATT
http://www.bioinf.man.ac.uk/dbbrowser/PRINTS/ PRINTS
http://blocks.fhcrc.org/ BLOCKS
http://blocks.fhcrc.org/blocks/blocks_search.html Block Searcher
http://bioinf.man.ac.uk/cgi-bin/dbbrowser/fingerPRINTScan/muppet/FPScan.cgi FTScan

http://www.bioinf.man.ac.uk/fingerPRINTScan/ FingerPRINTScan
http://www.sanger.ac.uk/Software/Pfam/index.shtml PFAM
http://hmmer.wustl.edu/ HMMER
http://www.ebi.ac.uk/interpro/ InterPro

Functional annotation
http://www.chem.qmw.ac.uk/iubmb/enzyme/ IUBMB Enzyme Nomenclature
http://us.expasy.org/enzyme/ Enzyme
http://www.genome.ad.jp/kegg/ KEGG
http://www.geneontology.org/ GO
http://mips.gsf.de/services/funcat FunCat

Localization prediction
http://psort.nibb.ac.jp/ PSORT, PSORTII, iPSORT
http://www.cbs.dtu.dk/services/TargetP/ TargetP
http://www.cbs.dtu.dk/services/SignalP-2.0/ SignalP
http://www.cbs.dtu.dk/services/ CBS Prediction Servers

Structure prediction
http://www.rcsb.org/pdb/ PDB
http://www.expasy.org/swissmod/SWISS-MODEL.html SWISS-MODEL
http://www-cryst.bioc.cam.ac.uk//cgi-bin/coda/pet.cgi PETRA
http://cgat.ukm.my/spores/Predictory/structure_prediction.html Structure Prediction
with Online Services
http://www.genesilico.pl/meta/

Prediction server
http://cubic.bioc.columbia.edu/predictprotein/index.html PredictProtein
http://mgd.ahc.umn.edu/panal / PANAL
http://metafam.ahc.umn.edu/ MetaFam
http://pedant.gsf.de/ Pedant
http://www.hgmp.mrc.ac.uk/Registered/Webapp/pix/ Pix

References

1. Eddy, S.R.: Computational genomics of noncoding RNA genes. *Cell* 2002;19:137-40.
2. Davuluri, R.V., Grose, I., Zhang, M.Q.: Computational identification of promoters and first exons in the human genome. *Nat Gen* 2001; 29:412-417.
3. Tabaska, J.E., Davuluri, R.V., Zhang, M.Q.: Identifiying the 3'terminal exon in human DNA. *Bioinformatics* 2001; 17:602-607.
4. Lowe, T. and Eddy, S.R.: tRNAscan-SE: a program for improved detection of transfer RNA genes in genomic sequence. *Nucleic Acid Res* 1997; 25:955-964.
5. Wuyts, J., Perrière, G., Van de Peer, Y.: The European ribosomal RNA database. *Nucleic Acid Res* 2004; 32:D101-D103.
6. Guigo, R., Agarwa, P., Abril, J.F., Burset, M., Fickett, J.W.: An Assessment of gene prediction accuracy in large DNA sequences. *Genome Res* 2000; 10:1631-1642.
7. Clote, P., Backofen, R.: *Computational Molecular Biology an Introduction.* Chichester, Wiley, 2000.
8. Zhang, M.Q.: Computational Prediction of Eukaryotic Protein-coding genes. *Nat Genet* 2002; 3:698-709.
9. Mathe, C., Sagot, M.F., Schiex, T., Rouze, P.: Current methods of gene prediction, their strengths and weaknesses. *Nucleic Acids Res* 2002; 30:4103-4117.
10. Frishman, D., Moronov, A., Mewes, H.W., Gelfand, M.: Combining diverse evidence for gene recognition in completely sequenced bacterial genomes. *Nucleic Acid Res* 1998; 26:2941-2947.
11. Lukashin, A.V. and Borodovsky, M.: GeneMark.hmm: new solutions for gene finding. *Nucleic Acid Res* 1998; 26:1107-1115.
12. Salzberg, S.L., Dlcher, A.L., Kasif, S., White, O.: Microbial gene identification using interpolated Markov models. *Nucleic Acid Res* 1998; 26:544-548.
13. Rogic, S., Mackworth, A.K., Ouellette, B.FF.: Evaluation of gene finding programs. *Genome Res* 2001; 11: 817-832.
14. Rogic, S., Ouellette, B.F.F., Mackworth, A.K.: Improving Gene Recognition Accuracy by Combining Predictions from Two Gene-Finding Programs. *Bioinformatics* 2002; 16:1034-1045.
15. Korf, I., Flicek, P., Duan, D., Brent, M.R.: Integrating genomic homology into gene structure prediction. *Bioinformatics* 2001; Suppl. 17:S140-S148.
16. Meyer, I.M. and Durbin, R.: Comparative ab initio prediction of gene structures using pair HMMs. *Bioinformatics* 2002; 18:1309-18.
17. Allen, J.E., Pertea, M., Salzberg, S.L.: Computational gene prediction using multiple sources of evidence. *Genome Res* 2004; 14:142-148.
18. Lewis, S.E., Searle, S.M., Harris, N., Gibson, M., Lyer, V., Richter, J., Wiel, C., Bayraktaroglir, L., Birney, E., Crosby, M.A., Kaminker, J.S., Matthews, B.B., Prochnik, S.E., Smithy, C.D., Tupy, J.L., Rubin, G.M., Misra, S., Mungall, C.J., Clamp, M.E.: Apollo: a sequence annotation editor. *Genome Biology* 2002; 3:research0082.1-0082.14.
19. Schoof, H., Ernst, R., Nazarov, V., Pfeifer, L., Mewes, H.W., Mayer, K.FX.: MIPS Arabidopsis thaliana database (MAtDB): an integrated biological knowledge resource for plant genomics. *Nucleic Acid Res* 2004; 32:D373-D376.

20. Burge, C. and Karlin, S.: Prediction of complete gene structures in human genomic DNA. *J Mol Biol* 1997; 268:78-94.
21. Kulp, D., Haussler, D., Reese, M.G., Eeckman, F.H.: A generalized Hidden Markov Model for the recognition of human genes in DNA. *Proc Int Conf Intell Syst Mol Biol* 1996; 134-142.
22. Salamov, A., Solovyev, V.: Ab initio gene finding in Drosophila genomic DNA. *Genome Res* 2000; 10:516-522.
23. Yeh, R.F., Lim, L.P., Burge, C.B.: Computational Inference of homologous gene structures in the human genome. *Genome Res* 2001; 11:803-816.
24. Wiehe, T., Gebauer-Jung, S., Mitchell-Olds, T., Guigo, R.: SGP-1: Prediction and Validation of Homologous Genes Based on Sequence Alignments. *Genome Res* 2001; 11:1574-1583.
25. Alexandersson, M., Cawley, S., Pachter, L.: SLAM: Cross-species gene finding and alignment with a generalized pair Hidden Markov Model. *Genome Res* 2003; 13:496-502.
26. Pertea, M., Lin, X., Salzberg, S.L.: GeneSplicer: a new computational method for splice site prediction. *Nucleic Acids Res* 2001; 29:1185-90.
27. Iseli, C., Jongeneel, V., Bucher, P.: ESTScan: a program for detecting, evaluating, and reconstructing potential coding regions in EST sequences. *Proc Int Conf Intell Syst Mol Biol* 1999; 138-48.
28. Usuka, J., Zhu, W., Brendel, V.: Optimal spliced alignment of homologous cDNA to a genomic DNA template. *Bioinformatics* 2000; 16:203-211.
29. Schlueter, S.D., Dong, Q., Brendel, V.: GeneSeqer@PlantGDB: gene structure prediction in plant genomes. *Nucleic Acid Res* 2003; 31:3597-3600.
30. Florea, L., Hartzell, G., Zhang, Z., Rubin, G.M., Miller, W.: A computer program for aligning a cDNA sequence with a genomic DNA sequence. *Genome Res* 1998; 8:967-974.
31. Birney, E., Thompson, J.D., Gibson, T.J.: PairWise and SearchWise: finding the optimal alignment in a simultaneous comparison of a protein profile against all DNA translation frames. *Nucleic Acid Res* 1996; 24:2730-2739.
32. Altschul, S.F., Gish, W., Miller, W., Myers, E.W. and Lipman, D.J.: Basic local alignment search tool. *J Mol Biol* 1990; 215:403-410.
33. Altschul, S.F., Madden, T.L., Schaeffer, A.A., Zhang, J., Zhang, Z., Miller, W., and Lipman, D.J.: Gapped BLAST and PSI-BLAST: a new generation of protein database search programs. *Nucleic Acids Res* 1997; 25:3389-3402.
34. Pearson, W.R.:Using the FASTA program to search protein and DNA sequence databases. *Methods Mol Biol* 1994; 24:307-31.
35. Pearson, W.R.: Flexible sequence similarity searching with the FASTA3 program package. *Methods Mol Biol* 2000; 132:185-219.
36. Boeckmann, B., Bairoch, A., Apweiler, R., Blatter, M., Estreicher, A., Gasteiger, E., Martin, M.J., Michoud, K., O'Donovan, C., Phan, I., Pilbout, S., and Schneider, M.: The Swiss-Prot protein knowledgebase and its supplement TrEMBL in 2003. *Nucleic Acids Res* 2003; 31: 365-370.
37. Apweiler, R., Bairoch, A., Wu, C.H., Barker, W.C., Boeckmann, B., Ferro, S.,

Gasteiger, E., Huang, H., Lopez, R., Magrane, M., Martin, M.J., Natale, D.A., O'Donovan, C., Redaschi, N. and Yeh, L.L.: UniProt: the Universal Protein knowledgebase. *Nucleic Acids Res* 2004; 32: D115-D119.

38. Wheeler, D.L., Church, D.M., Edgar, R., Federhen, S., Helmberg, W., Madden, T.L., Pontius, J.U., Schuler, G.D., Schriml, L.M., Sequeira, E., Suzek, T.O., Tatusova, T.A. and Wagner, L.: Database resources of the National Center for Biotechnology Information: update. *Nucleic Acids Res* 2004; 32(1):D35-40.

39. Boguski, M.S., Lowe, T.M. and Tolstoshev, C.M: dbEST-database for expressed sequence tags. *Nat Genet* 1993; (4):332-3.

40. Hulo, N., Sigrist, C.J.A, Le Saux, V., Langendijk-Genevaux, P.S., Bordoli, L., Gattiker, A., De Castro, E., Bucher, P. and Bairoch, A.: Recent improvements to the PROSITE database. *Nucleic Acids Res* 2004; 32: D134-D137.

41. Gattiker, A., Gasteiger, E. and Bairoch, A.: ScanProsite: a reference implementation of a PROSITE scanning tool. *Applied Bioinformatics* 2002; 1:107-108.

42. Jonassen, I., Collins, J.F. and Higgins, D.: Finding flexible patterns in unaligned protein sequences. *Protein Science* 1995; 4(8):1587-1595.

43. Jonassen, I.: Efficient discovery of conserved patterns using a pattern graph. *Comput Appl Biosci* 1997; 13(5):509-522.

44. Attwood, T.K., Bradley, P., Flower, D.R., Gaulton, A., Maudling, N., Mitchell, A.L., Moulton, G., Nordle, A., Paine, K., Taylor, P., Uddin, A. and Zygouri, C.: PRINTS and its automatic supplement, prePRINTS. *Nucleic Acids Res* 2003; 31: 400-402.

45. Henikoff, J.G., Greene, E.A., Pietrokovski, S., Henikoff, S.: Increased coverage of protein families with the blocks database servers. *Nucleic Acids Res* 2000; 28:228-230.

46. Henikoff, S., Henikoff, J.G. and Pietrokovski, S.: Blocks+: A non-redundant database of protein alignment blocks derived from multiple compilations. *Bioinformatics* 1999; 15(6):471-479.

47. Scordis, P., Flower, D.R., Attwood, T.K.: FingerPRINTScan: Intelligent searching of the PRINTS motif database. *Bioinformatics* 1999; 15(10): 799-806.

48. Bateman, A., Coin, L., Durbin, R., Finn, R.D., Hollich, V., Griffiths-Jones, S., Khanna, A., Marshall, M., Moxon, S., Sonnhammer, E.L.L., Studholme, D.J., Yeats, C. and Eddy, S.R.: The Pfam protein families database. *Nucleic Acids Res* 2004; 32: D138-D141.

49. Eddy, S.R.:Profile Hidden Markov Models. *Bioinformatics* 1998; 14:755-763

50. Mulder, N.J., Apweiler, R., Attwood, T.K., Bairoch, A., Barrell, D., Bateman, A., Binns, D., Biswas, M., Bradley, P., Bork, P., Bucher, P., Copley, R.R., Courcelle, E., Das, U., Durbin, R., Falquet, L., Fleischmann, W., Griffiths-Jones, S., Haft, D., Harte, N., Hulo, N., Kahn, D., Kanapin, A., Krestyaninova, M., Lopez, R., Letunic, I., Lonsdale, D., Silventoinen, V., Orchard, S.E., Pagni, M., Peyruc, D., Ponting, C.P., Selengut, J.D., Servant, F., Sigrist, C.J.A., Vaughan, R. and Zdobnov, E.M.: The InterPro Database, 2003 brings increased coverage and new features. *Nucleic Acids Res* 2003; 31: 315-318.

51. Zdobnov, E.M. and Apweiler, R.: InterProScan - an integration platform for the signature-recognition methods in InterPro. *Bioinformatics* 2001; 17:847- 848.
52. Bairoch, A.: The ENZYME database in 2000. *Nucleic Acids Res* 2000; 28: 304-305.
53. Ruepp, A., Zollner, A., Maier, D., Albermann, K., Hani, J., Mokrejs, M., Tetko, I., Guldener, U., Mannhaupt, G., Munsterkotter, M., Mewes, H.W. The FunCat, a functional annotation scheme for systematic classification of proteins from whole genomes. *Nucleic Acids Res.* 2004; 32(18): 5539-5545.
54. Gene Ontology Consortium: The Gene Ontology (GO) database and informatics resource. *Nucleic Acids Res* 2004; 32(1): D258-61.
55. Nakai, K. and Kanehisa, M.: Expert system for predicting protein localization sites in Gram-negative bacteria. *PROTEINS: Structure, Function, and Genetics* 1991; 11: 95-110.
56. Gardy, J.L., Spencer, C., Wang, K., Ester, M., Tusnady, G.E., Simon, I., Hua, S., deFays, K., Lambert, C., Nakai, K. and Brinkman, F.S.L.: PSORT-B: improving protein subcellular localization prediction for Gram-negative bacteria. *Nucleic Acids Res* 2003; 31(13): 3613-3617.
57. Horton, P. and Nakai, K.: PSORT: a program for detecting sorting signals in proteins and predicting their subcellular localization. *Trends Biochem Sci* 1999; 24:34-36.
58. Bannai, H., Tamada, Y., Maruyama, O., Nakai, K. and Miyano, S.: Extensive feature detection of N-terminal protein sorting signals. *Bioinformatics* 2002; 18: 298-305.
59. Emanuelsson, O., Nielsen, H., Brunak, S. and von Heijne, G.: Predicting subcellular localization of proteins based on their N-terminal amino acid sequence. *J Mol Biol* 2000; 300: 1005-1016.
60. Nielsen, H., Engelbrecht, J., Brunak, S. and von Heijne, G.: Identification of prokaryotic and eukaryotic signal peptides and prediction of their cleavage sites. *Protein engineering* 1997; 10: 1-6.
61. Schwede, T., Kopp, J., Guex, N. and Peitsch, M.C.: SWISS-MODEL: an automated protein homology-modeling server. *Nucleic Acids Res* 2003; 31: 3381-3385
62. Deane, C.M. and Blundell, T.L.: A novel exhaustive search algorithm for predicting the conformation of polypeptide segments in proteins. *Proteins: Struct Funct Genet* 2000; 40(1):135-144.
63. Rost, B. and Liu, J.: The PredictProtein server. *Nucleic Acids Res* 2003; 31(13): 3300-3304.
64. Silverstein, K.A.T., Kilian, A., Freeman, J.L. and Retzel, E.F.: PANAL: an integrated resource for Protein sequence ANALyis. *Bioinformatics* 2000; 16: 1157-1158.
65. Frishman, D., Mokrejs, M., Kosykh, D., Karstenmuller, G., Kolesov, G., Zubrzycki, I., Gruber, C., Geier, B., Kaps, A., Volz, A., Wagner, C., Fellenberg, M., Heumann, K., Mewes, H.: The Pedant genome database. *Nucleic Acids Res* 2003; 31: 207-211.

66. Henderson, J., Salzberg, S. and Fasman, K.H.: Finding genes in DNA with a Hidden Markov Model. *J Comp Biol* 1997; 4: 127-141.

Chapter 3

Bioinformatics of Gene Expression Analysis

M. Eisenacher

Contact: M. Eisenacher
Integrierte Funktionelle Genomik, Westfälische Wilhelms-Universität Münster,
von-Esmarch-Str. 56, D-48149 Münster, Germany
E-mail: eisenach@uni-muenster.de

3.1 Introduction

Massively parallel mRNA measurements using microarrays are performed for three main reasons: Firstly, the molecular characteristics (gene expression profile) of a tissue enable a "molecular diagnosis" of types and subtypes and based on that, a better prognostic statement or a better suited therapy. Secondly, known genes are being used for understanding the mechanisms of disease development (e.g. in cancer). Thirdly, functional characterization of unknown genes is possible through comparison with the behavior of known genes. In order to solve these three tasks, statistical or biomathematical methods are applied (e.g. classification, statistical tests, clustering).

Bioinformatics in a rigorous sense deals with algorithmic problems of biological issues. In the field of gene expression analysis the border between bioinformatics and biomathematics is rather blurred, as statistical analyses are often understood as bioinformatics. This may arise from the high-throughput character and the overwhelming size of experimental data, which is indeed a computational challenge. Nevertheless statistical analysis methods of expression data are not a topic of this chapter (see [1] for a detailed discussion of methods and software).

Gene expression measurements or "transcriptome analyses" with microarrays are always based on hybridization, which is the process of forming hybrids of mRNA and spatially fixed DNA molecules (probes) with complementary base sequence. The first part of this chapter deals with the choice of probes in a technology called "high-density oligonucleotide microarrays".

As very much data points are created in one experiment and due to the complex probe preparation process, questions on technical variance and reproducibility are becoming very important and will be discussed in the second part of this chapter.

3.2 Probe selection and condensing of microarray data

As the simplest choice for a probe molecule of a known target mRNA one may think about the complete complementary DNA molecule. But complete cDNA sequences cannot be used in practice, because of potential cross-hybridization due to closely related isoforms (similar mRNAs of other proteins) or produced by repetitive elements as part of the cDNA. In contrast, small excerpts of the complete sequence may not be specific to one gene, but be part of many transcripts, so that it cannot be decided from which transcript the hybridized material arose. On the other hand a probe position may be located in a region that is not part of all possible transcripts (e.g. alternative polyadenylation, alternative splicing, alternative promoter usage). An additional aspect is, that hybridization takes place under uniform experimental conditions (e.g. temperature, pH) for each probe / target pair, so that probe selection should optimize the global hybridization behavior with respect to probe length and melting point of the hybrids.

The extraction of suitable probes out of the variety of sequence information resources for the detection of mRNA abundance is an interesting field of bioinformatics. Normally, on spotted arrays one long library cDNA or PCR product is used for the measurement of one transcript and on high-density oligonucleotide arrays multiple probes of 25 base pairs long are used and their measurements assembled to one abundance value. Of course the probe selection process differs between these two platforms. Probe selection strategies as sketched above are generally discussed in [2]. An example of a web resource with links to different probe design tools in[3].

In the following, at first a short recapitulation of gene databases is offered. Then the probe selection method for current Affymetrix high-density oligonucleotide arrays ("GeneChips®") is discussed and after that the process of total signal assembly ("condensing") is reviewed.

3.2.1 Gene Databases

In the beginning of the genomic era the sequences of analyzed genes were published in text form in journal articles (as for example in [4] for the yeast gene iso-2-cytochrome C). When electronic resources became available, known sequences were transferred into databases together with the corresponding article reference (see entry "V01299" in GenBank[5] for the above mentioned yeast gene). With ongoing sequencing efforts the process of publishing a sequence changed to reversed order: Nowadays an identified sequence is submitted to a database, gets a unique identifier and is then cited in a publication. In the following summarizing facts about sequence databases of the "National Center for Biotechnology Information"[6] are exemplarily presented, since they are relevant for later sections.

GenBank

The first published presentation of GenBank was done in[7], where it was characterized as a computerized storage of all published nucleotide sequences (DNA and RNA) with five million bases in about 6000 sequences from 4500 publications. Nowadays GenBank[8] contains more than 37 billion bases from more than 32 million individual sequences.

A submitted sequence first of all gets a unique identifier (ACCESSION number). The format of an ACCESSION number is "one letter followed by five digits", e.g. U12345, or "two letters followed by six digits", e.g. AY123456 for nucleotide sequences. As all versions of an corrected or edited entry are archived, a VERSION field is part of an entry with the format "ACCESSION.<version>". Each entry has a unique identifier GI.

GenBank is a very voluminous and open sequence repository, but in return very little controlled.

dbEST

In most cases transparent to the web user, GenBank is partitioned into different parts, the so-called divisions. In the past, these parts mainly depended on species origin (e.g. PRI: primate sequences). In[9] a special EST division of GenBank is presented, which contains "expressed sequence tags". ESTs are short (around 300-

500 bp), single-pass sequence reads from mRNA (cDNA) and represent expressed genes (transcripts) in an examined tissue / material. Naturally the same sequence may be submitted dozens or hundreds of times, found in different material. The EST division of GenBank (dbEST) has a separate web home page[10].

UniGene

As many dbEST entries of varying annotation quality may exist for the same transcript, another database called UniGene was firstly presented in [11] as an abstraction of dbEST. One UniGene entry is a grouping of GenBank sequences and can be seen as a non-redundant set of gene-oriented EST clusters. It contains information about the common sequence, the EST entries, that build the cluster, the tissues, in which the transcript is expressed and the "best" similarities to proteins of other species. UniGene is part of the Entrez web portal[12].

RefSeq

As GenBank is an archive of all submitted sequences ever and the submitter is responsible for correct and detailed annotation without any review, the necessity for standardized reference sequence entries became apparent. Therefore, the RefSeq database was created (current status:[13]; website:[14]). An entry contains a best-characterized sequence, a consistent description, uses standard nomenclature and is always corrigible. Different types of review stages exist; the best one, "Reviewed", implies "manual" review by a biologist. RefSeq accession numbers consist of a two-letter prefix (first one: N or X), followed by an underscore and a number, as for example "NM_001005738".

3.2.2 Probe selection

The probe selection process of Affymetrix high-density oligonucleotide mircoarrays can be divided into three major parts (see [15]):

- **Sequence selection**: Which sequences from the public databases should be used as representatives for existing transcripts? Because of many equal or similar EST sequences these databases are highly redundant. Probe sets should be placed in the region next to the 3' or polyadenylated (PolyA) end of the sequence, because laboratory preparation steps *reverse transcription* (RT) and *in vitro transcription* (IVT) run from 3' to 5' and are the less successful in producing efficiently detectable target, the longer they run. As a result for different transcript variants (e.g. alternative PolyA sites) different sequences should be chosen for a reasonable probe selection.

- **Probe region selection**: Once a sequence is determined, the region for probe selection must be specified. As translated regions may contain motives, which are part of many other proteins, the probe selection region should be ideally placed in the 3' untranslated region (3' UTR). If such an ideal placement of the probe region cannot be done, alternative strategies have to be applied dependent on the completeness of annotated information.

• **Probe set selection**: A set of oligonucleotides is chosen from the probe region to ensure <u>uniqueness</u> of each probe (no cross hybridization of other sample material) and hybridization characteristics, which optimize <u>probe binding affinity</u> and <u>linearity of signal changes</u> in response to target variations.

In the probe selection processes of former GeneChip generations main problems existed with weakly annotated, poorly characterized, seldom expressed sequences. In some cases it remains unknown, in which direction a sequence is oriented, whether it contains coding regions or where a PolyA site is placed. In the design of the HG-U133 array for the first time available genomic sequence information was used to deal with some of that problems. In addition, primary sequence and annotation information was combined from a large variety of public databases.

Firstly a <u>basic pool of sequences is collected</u> from the utilized data sources. The raw sequence together with sequence Meta information as for example clone identifiers and read directions are considered.

Then alignment of all sequences to the genome is calculated to <u>trim low quality bases</u> from ESTs. In addition, the genomic alignment can confirm or newly identify sequence orientation through consensus splice sites. Because of the above mentioned fact about efficiency of RT and IVT, probe regions should be placed as near as possible next to a PolyA site (generally 600 bp upstream), so great effort is done to identify such sites. For this reason untrimmed, primary sequence information was utilized and helped significantly because PolyA or PolyT parts are often deleted before submission to public databases.

Once the cleaned pool of sequences is created, the aim now is to <u>find largest subsets of sequences ("design cluster")</u>, so that representative sequences can consistently be derived that result in reasonable probe regions. UniGene clusters (from UniGene Build U133) are used as "seed clusters". Because the seed clusters may contain sequences, which do not consistently lead to reasonable probe regions, they are divided into smaller clusters by three different processes of subclustering: 1) <u>genomic-based subclustering:</u> all sequences in a subcluster must not lie in different contigs (e.g. chromosomes); 2) <u>sequence-based subclustering:</u> all sequences in a subcluster must be 75% identical when a consensus is called (ambiguous and polymorphic bases dropped); 3) <u>orientation-based subclustering:</u> CDS annotation, PolyA identification or flanking introns of genomic alignment are used to build subclusters of "sense" and "anti-sense" oriented sequences ("unknowns" are put into the larger subcluster).

The resulting design clusters typically represent one transcript variant (eventually with several alternative polyadenylation sites). Based on the orientation call, the 3' end of a cluster is determined (in clusters of unknown orientation, probe regions are selected on both ends of the sequence, for example the design cluster resulting in probe set 216043_x_at). Potential <u>transcript ends</u> are identified (*i*) by the 3' end of a full length member mRNA sequence, (*ii*) by a set of 8 or more agreeing EST ends or (*iii*) by the end of the consensus sequence (see figure 3-1).

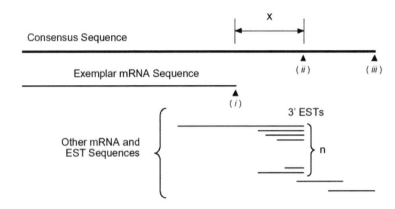

Figure 3-1 (slightly modified from [15]). *Probe region selection in a design cluster.*

As multiple transcript ends may be identified, multiple probe selection regions may exist for one design cluster:

If a full-length sequence is part of the cluster and it has a 3' UTR, then probes are selected from the region 600 bp before (i). If n ≥ 8 ESTs agree for an additional transcript end and the distance to (*i*) is large enough (x > 400), then a probe region 600 bp before (*ii*) is selected (if n < 8 or x < 400, this region (*ii*) is not selected).

If the potential full-length sequence has no 3' UTR, than (*i*) is not used as probe region. Region (*ii*) is only used, if n ≥ 8. Otherwise, a probe region 600 bp before (*iii*) is used.

If a full-length mRNA identified one transcript end, then this mRNA sequence is used as the representative "<u>exemplar</u>" sequence for probe set selection. For all other transcript ends the "<u>consensus</u>" sequence is used.

The finally derived design clusters are labeled with internal identifiers of the format "Hs.*xxxx.y*", for example Hs.79732.1 for a subcluster of the UniGene (build 133) seed cluster Hs.79732 representing the fibulin 1 gene (Hs.79732 unfortunately does not exist in the current UniGene build 177; as the most similar cluster Hs.24601 is given). Representing sequences of potential transcript ends can be exemplar sequences (format of internal identifier: "g*xxxxxxx*", for example g5922008) or consensus sequences (starting with "Hs." as above). The internal identifiers of these representing sequences are available for each probe set in the NetAffx Analysis Center [16] for registered users, as annotation column "Transcript ID (Array Design)".

As an example, the resulting probe sets of transcripts annotated with "fibulin 1" are used here (as in [15]): 201787_at, 202994_s_at, 202995_s_at, 207834_at, 207835_at. In figure 3-2 all design clusters are schematically shown, pairwise aligned to the seed sequence Hs.79732 (first line). On the left the design cluster labels are given, together with the internal identifiers of the exemplar and consensus sequences. A potential full-length sequence is shown (Hs.900486.556) in this alignment, which is not part of UniGene. Probe sets 201787_at, 202995_s_at, 207834_at and 207835_at are (in this order) represented by exemplar sequences annotated in GenBank with "transcript variants" C, D, B and A. 202994_s_at is derived from a consensus sequence of a cluster containing ESTs, that give hints to an alternative PolyA site.

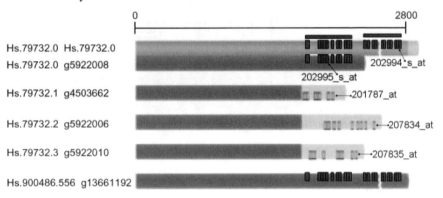

Figure 3-2 (slightly modified from [15]). "fibulin 1" design clusters, exemplar / consensus sequences and probe sets.

Aligned to the genome, the exemplar / consensus sequences of the fibulin 1 probe sets of the HG-U133_Plus_2.0 array are located as visualized with the UCSC genome browser[17] in Figure 3-3. At the top of the figure the chromosome overview gives the position of the visualized area. Solid blocks (often condensed to vertical lines) visualize exons. The first group of lines visualizes exemplar / consensus sequences, the second and third group visualize known genes, especially RefSeq genes, here manually complemented by RefSeq isoform annotation.

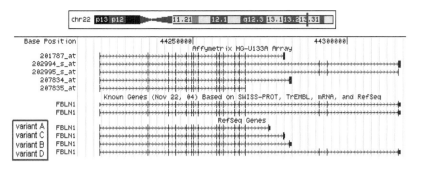

Figure 3-3. *Alignment of fibulin 1 probe sets to the genome.*

After the probe regions are selected, the actual probes have to be determined. This process is described in detail in [18]. There a thermodynamic model is formulated for the prediction of probe response. In this model target concentrations are considered together with some unfavorable interactions like consecutive hairpins, nonconsecutive hairpins and G quartets (hydrogen-bonded G tetraplexes).

With hybridization experiments on two specially designed GeneChips, results of varying concentrations for some known targets covering a 4000fold range are measured. These data are used to establish the prediction model and to test model quality via cross-validation. Two different model equations have been used, a linear equation for low-affinity probes and a sigmoid equation for high-affinity probes (like probes with high GC content), which takes into account the non-linearity of chemical saturation in an empirical manner.

A figure in [18] visualizes the correlation coefficient ranges between predicted and measured intensities of all measured transcripts for each target concentration. The averages of these correlation coefficients consistently lie above 0.8.

The combination of the modeled response metric with criteria concerning uniqueness (avoid cross-hybridization with other known expressed sequences in the genomic background) and independence (avoid probes with overlapped sequences, which may be vulnerable to similar systematic errors) enabled the selection of optimal probe sets in a systematic and large-scale manner.

As an example of resulting probes in figure 3-4 the three most right probes of probe set 202994_s_at are shown, aligned to the corresponding probe region.

```
...gaaatagaggaaaatcccttggtaaagacacagcctgttaggctcgtgtgggcctccagtatgttcaccaggggaa
                 atcccttggtaaagacacagcctgt
                          acagcctgttaggctcgtgtgggcc
                                       gcctccagtatgttcaccaggggaa
```

Figure 3-4. *The three right-most probes of probe set 202994_s_at; underlined: mismatch positions.*

From a genomic view, the probe region (and therefore the probe set) can be spanned over intron gaps, as visible in figure 3-5 (produced with the "Internal Genome Browser, IGB", available via the NetAffx website).

Figure 3-5. Probe sets (small black marks) may be spread over different exons (grey).

Finally, for each of the above selected "Perfect Match" probes, "Mismatch" probes are created by switching the base in the middle to its complementary base.

3.2.3 Condensing of intensities to a *Signal* value

The special architecture of Affymetrix high-density oligonucleotide microarrays (summed up in Figure 3-6) with intensity values for Perfect Match / Mismatch probe pairs requires a special tailored algorithm, that calculates the actual value, that should be proportional to the mRNA concentration in the sample. In a former algorithm, called "empirical algorithm", this value was called "Average Difference" and could get negative, which unfortunately had no biological interpretation. Now, the algorithm is called "statistical algorithm" and the positive abundance value is called "*Signal*".

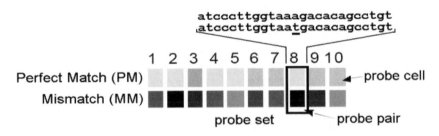

Figure 3-6 (modified from [19]*). GeneChip design and nomenclature.*

In addition to the *Signal* value, some other criteria are calculated, as for example a "Detection p-value" and a "Detection Call", which enable the judgment, whether the measured value is reliable. A detailed algorithm explanation can be found in [19]; here a short summary of the major steps of *Signal* value calculation is given.

Background subtraction

The array is divided into K (default: K=16) spatial subsections (zones). Control and masked probe cells are not used in the calculation. Of the remaining cells the 2% with lowest intensities are chosen. The average intensity of them is the background for that zone (bZ_k) and the standard deviation of them is the noise for that zone (nZ_k). The aim is now, to correct the intensity of each probe cell by subtracting the zone background of that cell. To ensure a smooth transition between zones, the background of the other zones is subtracted, too, but weighted the less important, the greater the distance between probe cell and zone center is. The overall background to be subtracted is abbreviated $b(x,y)$ for a probe cell with coordinates (x,y). Analogously, $n(x,y)$ is a weighted sum of the zone noises. It is used to avoid negative values resulting from background subtraction.

The background subtraction may be sketched as follows:

$$A(x, y) = \max \left(I(x, y) - b(x, y), 0.5 \cdot n(x, y) \right)$$

where $I(x,y)$ is the raw intensity of probe cell (x,y). Illustratively: The background is subtracted from the raw intensity, but the result must not be smaller than half of the noise for that probe cell.

Adjustment of Perfect Match intensity

The Mismatch (MM) probe cells are constructed to detect unspecific hybridization. Therefore subtraction of Mismatch from Perfect Match (PM) value has to be performed for every probe pair. Unfortunately it is possible, that MM intensity is larger than PM intensity. This is an unexpected case and may arise for example from unintentional specific hybridization of sample material to MM (e.g. due to mutations). Of course, in this case the MM cannot be used in the intended way, but is replaced by a so-called Ideal Mismatch (IM). Illustratively spoken is that IM forced to be smaller than the corresponding PM, approximately by a representative ratio between PM and MM of all probe pairs of the probe set:

$$IM_j = \begin{cases} MM_j, & MM_j < PM_j \\[2ex] \dfrac{PM_j}{\text{representative ratio}}, & MM_j \geq PM_j \end{cases}$$

for probe pair j. For further details of the above case differentiation and the calculation of the representative ratio see [19].

After that, the difference $V_j = PM_j - IM_j$ is calculated. If this is too small ($< 2^{-20}$), it is replaced by 2^{-20} for numerical stabilization.

Calculation of robust mean

For a probe set i now values V_j represent the measurements of different probes for one transcript. Ideally they would measure the same value, which is proportional to the sample mRNA concentration. In reality a bunch of reasons are imaginable, which make these values vary between the probe pairs. Nonetheless the majority of probe pairs should point to a similar value. To account for outliers, neither the mean nor the median is used as the final probe set *Signal* value, but a robust mean calculated by an algorithm named Tukey's Biweight algorithm. In this algorithm data points have the less influence on the result, the further they lie from the median.

The result of the algorithm started on the V_j of all probe pairs j of a probe set i is the (unscaled) *Signal* value of i.

Standard scaling

Standard scaling shifts the mean of the value range of all probe sets to a target value by applying a linear factor — the so-called scaling factor — to all *Signal* values. The 2% lowest and highest *Signal* values are excluded from the mean calculation, therefore it is also called "trimmed mean". The default target value is 500, but can be changed to other values.

Some publications discuss improvements of the condensing: In [20] a model-based process is presented, which considers several chips of a data set in contrast to only one single experiment. A method especially for paired chips (e.g. "before / after" replicates) is developed in [21]. Comparisons between different condensing algorithms are presented in [22] or [23].

3.3 Variance considerations

When performing high-density oligonucleotide experiments great efforts must be made to ensure reproducible laboratory conditions during the complex probe preparation consisting of steps "material extraction", "mRNA isolation", "reverse transcription (RT)", "in vitro transcription (IVT)", "fragmentation", "filling of chip", "hybridization" and others.

In contrast to other expression measurements, not only one, but very many data points are affected and small changes in laboratory conditions have an unforeseeable more or less important influence on *Signal* values. Even worse, this influence can differ between probe sets in its magnitude or in its direction. Many sources of influence are imaginable and because of the expensiveness of repeated experiments (replicates) the type and behavior of variance between experiments dependent on different influence factors is an important research field.

In [24] several quality criteria are assessed to characterize variance sources on own and publicly available data sets. In the following sections, some considerations about possible influence factors are summed up.

3.3.1 Biological and technical influence factors

Variance is used in [24] as a synonym for the quantitative effects of factors, which influence the *Signal* value. Such an influence can arise from diverse sources. The roughest differentiation maybe, whether a change in *Signal* value is caused by reasons laying inherently in the "biology" of the sample material, or by reasons stemming from the measurement technology itself ("biological" vs. "technical" variance). Biological variance with this definition is then the effect, for that the experiment is actually designed. Of course, there exist as many biological variance sources as biological phenomena, some of which are for example: regulation of development, reaction to a drug, alterations as reason or effect of disease, position in cell cycle. Technical variance may have different reasons in different steps of laboratory workflow. The aim is to avoid it completely or if that is not possible in reality — to reduce it to a minimum. Figure 3-7 gives an idea of variance sources.

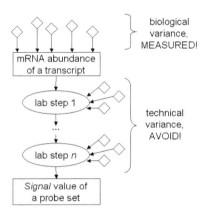

Figure 3-7. *Sources of variance.*

The total technical variance may be modeled like:

$$Signal(\text{probe set}_i) = (f_i \circ var_i^{\text{tech}}) \; c_{\text{sample}}(\text{mRNA}_i),$$

where f_i is the (ideally linear) measurement function, var_i^{tech} is the "disturbing" variance function and c_{sample} is the concentration of an mRNA$_i$. The first measurable impact of technical variance does not occur in the *Signal* value, but in the intensity measurement of a probe cell. Therefore another modeling possibility is:

$$intensity(\text{probe cell}_j) = (g_j \circ var_j'^{\text{tech}}) \; c_{\text{sample}}(\text{oligo}_j),$$

where g_j is the measurement function, $\text{var}_j{'}^{\text{tech}}$ the variance function and c_{sample} the concentration of oligonucleotide j in the processed material. The condensing algorithm may be formulated as a function *cond*:

Signal (probe set$_i$) = *cond* (CEL, *intensity* (probe cell$_k$), *intensity* (probe cell$_l$) | ∀ (PM,MM) probe pairs (k, l)),

where CEL are all global chip parameters that influence the algorithm (like background and noise) and (k,l) enumerate all probe pairs of probe set i. From the complex nature of the condensing algorithm follows, that $\text{var}_j^{\text{tech}}$ could be expressed with functions $\text{var}_j{'}^{\text{tech}}$ only in a nontrivial manner. Variance categories are for example Bias, Noise or local variance, which describe a reproducible, random or neighboring character of variance.

3.3.2 Behavior of technical variance

To get an idea of technical variance, the majority of probe sets is not useful, as they are potentially changed by biological variance. They could only be used in a data set consisting of technical replicates, which are usually too few to be able to draw conclusions.

Quality criteria like background and noise (given in a separate quality report) are more independent from experimental design and are used in [24] to describe technical variance behavior. It is shown, that they exhibit some variability within data sets, even within experimental groups (biological replicates) and that their averages may significantly differ between data sets.

Similar observations are made for some special designed probe sets: Hybridization Controls and Housekeeping Controls. The former measure transcripts added to the processed sample in known concentrations, and the latter measure transcripts already present in the sample material, for which no or only minor regulation is assumed (GAPDH and β-Actin). In addition, with separate probe sets for 3' and 5' regions of the Housekeeping transcripts problems with the RT and IVT steps of probe preparation can be identified. The 3'/5' ratio is ideally 1, indicating no *Signal* difference between 3' and 5' region, and greater than 1, indicating a lower *Signal* of the 5' probe.

The most interesting conclusions can be drawn from data sets, which contain technical replicates. These are replicates, which replicate within the probe preparation process. Data sets of this design are very seldom. In [24] a data set is used, replicating just before the step "filling of chip". Unfortunately, replicates are of Mu11Ksub<u>A</u> and Mu11Ksub<u>B</u> array variants, but anyway usable, because Housekeeping Control probe sets are identical. Average Signal values from different probe regions show a rather large variability within the whole data set (34 A and 34 B experiments), but drawing a scatter plot of A vs. corresponding B value reveals good correlation (see figure 3-8 for ß-Actin with a correlation coefficient r=0.96). As this is a consistent observation and reproducible with several other quality criteria, together with the replicating design, it can be concluded that the major fraction of technical variance is not located in steps after.

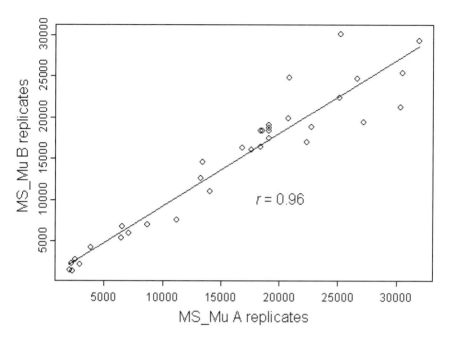

Figure 3-8. *Correlation of ß-Actin Signal values.*

To elucidate the nature of technical variance, a finer-grained design with technical replicates additionally after steps "isolation", "RT" and "IVT" would be eligible, but it is an expensive and a challenging task, because of necessary protocol modifications due to larger masses or concentrations of material. [24] attempts to characterize technical variance sources by studying global changes on probe cell intensity level. Differing intensity distributions hint to unspecific hybridization as issue of variance, but the major source remains undiscovered. "Extraction/Isolation", "RT", "IVT" or "fragmentation" steps are major candidates for important technical variance sources.

3.4 Conclusions

Expression studies of microarrays are widely used and offer great potentials in many fields of biology and medicine. In interpreting the results it is crucial to have a deep understanding of their design (e.g. in the presence of multiple probe sets for one gene).

The technical development of course does not stop at the current stage. Using the genome sequence and a modified protocol, microarrays will become possible, which can distinguish between different splice variants or transcripts with different promoter starts.

The sketched work on technical variance is an important step to make meta-analyses on large expression databases reasonable. Meta-analyses on publicly available expression data — analogous to meta-analyses on publicly available sequence data like BLAST — may give insights about gene expression behavior on a global scale.

Acknowledgments

The author wants to thank Thomas Vogl from the Department of Experimental Dermatology for collaboration.

References

1. Parmigiani, G., Garrett, E.S., Irizarry, R.A., Zeger, S.L. (Eds.), *The Analysis of Gene Expression Data - Methods and Software*, Springer-Verlag (2003), ISBN: 0-387-95577-1.
2. Tomiuk, S., Hofmann, K. (2001) Microarray probe selection strategies, *Brief. Bioinform.*, 2, 329-340.
3. Microarray Software Comparison - Probe / Primer Design Software (*http://ihome.cuhk.edu.hk/%7Eb400559/arraysoft_probe.html*, last visited: 2005-01-11)
4. Montgomery, D.L., Leung, D.W., Smith, M., Shalit, P., Faye, G., Hall, B.D. (1980) Isolation and sequence of the gene for iso-2-cytochrome c in Saccharomyces cerevisiae, *Proc. Natl. Acad. Sci.U.S.A*, 77, 541-545.
5. Genbank (*http://www.ncbi.nlm.nih.gov/Genbank/index.html*, last visited: 2004-12-29)
6. National Center for Biotechnology Information - Website (*http://www.ncbi.nlm.nih.gov/*, last visited: 2004-12-30)
7. Burks, C., Fickett, J.W., Goad, W.B., Kanehisa, M., Lewitter, F.I., Rindone, W.P., Swindell, C.D., Tung CS, Bilofsky HS (1985) The GenBank nucleic acid sequence database, *Comput. Appl. Biosci.*, 1, 225-233.
8. National Center for Biotechnology Information - GenBank (*http://www.ncbi.nlm.nih.gov/Genbank/index.html*, last visited: 2004-12-30)
9. Boguski, M.S., Lowe, T.M., Tolstoshev, C.M. (1993) dbEST—database for "expressed sequence tags", *Nat.Genet.*, 4, 332-333.
10. National Center for Biotechnology Information - dbEST (*http://www.ncbi.nlm.nih.gov/dbEST/index.html*, last visited: 2004-12-30)
11. Boguski, M.S., Schuler, G.D. (1995) ESTablishing a human transcript map, *Nat.Genet.*, 10, 369-371.
12. National Center for Biotechnology Information - UniGene (*http://www.ncbi.nlm.nih.gov/entrez/query.fcgi?db=unigene*, last visited: 2004-12-30)
13. Pruitt, K.D., Tatusova, T., Maglott, D.R. (2003) NCBI Reference Sequence project: update and current status, *Nucleic Acids Res.*, 31, 34-37.
14. National Center for Biotechnology Information - RefSeq (*http://www.ncbi.nlm.nih.gov/RefSeq/*, last visited: 2004-12-30).
15. Affymetrix, Inc. (*Technical Report*): Array Design for the GeneChip Human Genome U133 Set, (2001).
16. Affymetrix Website - NetAffx Analysis Center (*http://www.affymetrix.com/analysis/index.affx*, last visited: 2005-01-11)
17. Genome Browser - University of California, Santa Cruz (UCSC) (*http://genome.ucsc.edu/cgi-bin/hgGateway*, last visited: 2004-12-29)
18. Mei, R., Hubbell, E., Bekiranov, S., Mittmann, M., Christians, F.C., Shen, M.M., Lu, G., Fang, J., Liu, W.M., Ryder, T., Kaplan, P., Kulp, D., Webster, T.A. (2003) Probe selection for high-density oligonucleotide arrays, *Proc. Natl. Acad. Sci.U.S.A.*, 100, 11237-11242.

19. Affymetrix, Inc. (*Technical Report*): Statistical Algorithms Description Document, (2002).
20. Li, C., Wong, W.H. (2001) Model-based analysis of oligonucleotide arrays: expression index computation and outlier detection, *Proc. Natl. Acad. Sci. U.S.A.*, 98, 31-36.
21. Naef, F., Hacker, C.R., Patil, N., Magnasco, M. (2002) Empirical characterization of the expression ratio noise structure in high-density oligonucleotide arrays, *Genome Biol.*, 3, RESEARCH0018.
22. Rajagopalan, D. (2003) A comparison of statistical methods for analysis of high density oligonucleotide array data, *Bioinformatics.*, 19, 1469-1476.
23. Zhou, Y., Abagyan, R. (2003) Algorithms for high-density oligonucleotide array, *Curr. Opin. Drug Discov. Devel.*, 6, 339-345.
24. Eisenacher, M., Funke, H., Vogl, T., Cichon, C., Riehemann, K., Sorg, C., Köpcke, W., Intensity-based Variance Considerations and Scaling of High-Density Oligonucleotide Microarray Expression Data (manuscript submitted).

Chapter 4

An Introduction to EMBOSS — the free born son of GCG

Richard P. Grant

Contact: Richard P. Grant
Medical Research Council Laboratory of Molecular Biology, Cambridge, UK
E-mail: rpg@mrc-lmb.cam.ac.uk

4.1 What is Emboss?

EMBOSS is a free Open Source sequence analysis package. It was developed especially for the needs of the molecular biology user community[1].

4.1.1 What does it do?

As a software package, or suite, EMBOSS consists of over 150 individual programs[2]. These are run independently — from a script, a graphical interface, or explicitly from the command line — but act together to provide an integrated whole. There are also several additional programs which, while not strictly part of EMBOSS for licensing reasons, function in a similar manner[3]. EMBOSS is truly modular; new programs and utilities can be added without affecting the rest of the package.

The software can accept input data in a bewildering variety of formats[4], and allows for retrieval of sequence data from local sources as well as transparent retrieval from remote databases over the internet. Similarly, several output formats are available[5].

The programs that are part of the EMBOSS package can do just about anything one would wish to do with sequence data. They are organized within the package into about thirty groups[6] according to function. For example, there are groups of programs for creating, analysing and editing multiple sequence alignments, for prediction of two dimensional structure of nucleic acids and proteins, for phylogeny, for pattern recognition, and for creating and indexing databases.

4.1.2 Why does it do it?

In 1982 the Genetics Computer Group (GCG) was created as a service to the Department of Genetics at the University of Wisconsin. Their primary product was the Wisconsin suite of molecular biology tools, commonly known as just 'GCG'. This package became popular and well-respected in the academic community. Part of the package's strength was that the source code of the libraries it used was made available, and the programs could be verified algorithmically and adapted as necessary. Moreover, new programs which used the libraries could be written. Thus in 1988 was born the EGCG (Extended GCG) package, which began life at EMBL in Heidelberg, Germany. It originally provided a small collection of programs to support EMBL's research activities, particularly automated DNA sequencing. At the peak of its popularity EGCG was used at around 150 sites, including the Sanger Centre[7] and the Rosalind Franklin Centre for Genomics Research (RFCGR; formerly the Human Genome Mapping Project (HGMP) Resource Centre)[8], and by more than 10,000 users of EMBnet[1] national services.

However, in 1990 GCG became a private company, and ultimately was bought by Oxford Molecular Group in 1997[9]. When the company stopped releasing the GCG package's source code it was no longer possible to distribute source code which used the GCG libraries, and it became difficult to distribute binaries (compiled applications). Academics found themselves unable to build upon the GCG base package. More worryingly, the unavailability of source code was not good for true, 'open' science, as algorithms could not be checked and modified programs - perhaps more relevant to real world requirements - could not be written. Readers familiar with Phil Zimmermann's Pretty Good Privacy (PGP[10]) package will no doubt note parallels.

There are other commercial, standalone packages, aside from GCG, that duplicate some of the sequence analysis functions, but these tend to be prohibitively expensive both for the casual user and for departments who wish to roll out a site-wide solution. Additionally, the source code and libraries for these programs are surrounded by intellectual property barbed wire and can not be modified or improved by those on the 'frontline'.

Don Gilbert maintains a list of free software by many authors at <http://iubio.bio.indiana.edu/soft/molbio/Listings.html> but it is tedious to examine and install each of the individual programs, of frankly variable quality. The industrial strength yet free to academics Staden package (see <http://www.mrc-lmb.cam.ac.uk/pubseq/>) for sequence data manipulation and analysis has recently had its funding terminated, and its future is unclear.

Fortunately for open science (and academics' budgets!), former EGCG developers and others designed a totally new generation of sequence analysis software that resulted in the EMBOSS package. From the outset, the EMBOSS package was intended to be free, both in the sense of 'free beer' (i.e. it costs nothing to the user) and of 'free speech' (i.e. it can be modified and added to without complicated licencing agreements). As described on the EMBOSS website, it 'breaks the historical trend towards commercial software packages'[11]. The project proper was initiated in 1997 by Alan Bleasby and Peter Rice and EMBOSS version 1.0.0 was released on 15 July, 2000.

4.1.3 Is it for me?

Experienced computer users have a right to be somewhat suspicious of free or shareware software, as it often does not have the 'polish' nor the functionality of commercially published software. EMBOSS however is fully-featured and is a more than adequate replacement for the GCG/EGCG package. Indeed, the Oxford University Bioinformatics Centre[12], which has provided access to GCG since the early days, used to 'highly encourages' its users to try EMBOSS. Moreover, it is used extensively in production environments rather than being a research curiosity, with over 14,000 downloads in the first three years of its public release[11]. Many of these downloads have been for site-wide installations, so the number of individual users is likely to be much higher.

The package is mature (version 2.10.0 was released in February 2005) and stable. The programs are easy to use and come with extensive documentation[13]. New programs are tested extensively before being released. The size of the user base, coupled with effective user feedback mechanisms, ensures that any bugs that do make it into the wild are rapidly identified and fixed promptly. Similarly, suggestions for improvements and new features can be assessed and incorporated, something that is very rare with commercial packages. Indeed, the community aspect and open model of the EMBOSS project ensures a much closer relationship between user and developer, so that everyone — not merely a company's shareholders — benefits much more directly from user feedback. EMBOSS is now fully accepted as part of the Open Source community, being hosted by SourceForge.net <http://sourceforge.net>, the world's largest Open Source development website and software repository.

The most straightforward way to use the programs is to call them from a command line, but graphical interfaces are available and more are being developed ([14] and see below). Other advantages to the user include handling of all sequence and many sequence and alignment formats and integration with other publically available packages (e.g. Staden[15]).

From a developer's point of view, EMBOSS provides a properly constructed toolkit for creating robust bioinformatics applications or pipelines. Technically, it is free of arbitrary size limits — memory management in C is handled by the system and memory for sequences or matrices is allocated dynamically. It contains library functions for general string handling, pattern-matching, sorting, iteration and extremely fast indexing. There are library functions for all common sequence analysis tasks and a consistent API for interface designers. Extensive libraries are included with the package, and are covered by the GNU Lesser General Public Licence[16].

Finally in this section, the software runs on just about any flavour of UNIX[17,18]. Until quite recently this was somewhat limiting for 'ordinary' users who did not have access to an appropriate system at their site (the HGMP used to offer accounts and access to GCG, etc., but this service is being withdrawn with the closure of the RFCGR research and Bioinformatics divisions. Other freely accessible shell accounts remain available[19]). However, the latest version of the popular Macintosh operating system[20] is based on UNIX and is proving to be a capable platform for bioinformatics applications[21], including the EMBOSS package. This it is eminently possible to run EMBOSS on your Mac (at home, even!), without access to a mainframe class computer.

Oh, and did we mention that it is free?

4.2 Using EMBOSS

4.2.1. Downloading & Installing

The complete EMBOSS package can be downloaded from ftp://ftp.uk.embnet.org/pub/EMBOSS/. The installer package comes with detailed compilation and build instructions. Macintosh users can download and build the package (or install binaries) using Fink[22]. The entire package, including Jemboss (a graphical interface to EMBOSS), weighs in at around 13 MB, which is eminently reasonable even over a dial-up connection.

Note for Mac OS X users:

Before installing EMBOSS on your own machine, you will need to install the Developer Tools CD, which is freely available as a (somewhat large, admittedly) download from Apple. You will need to do this whether you are using the EMBOSS installer package or Fink. Additionally, you will need to install X11 and the X11 SDK. EMBOSS now works with the current (March 2005) version of Apple's X11 that shipped with Mac OS 10.3 (Panther). This is a far superior version of X11 and integrates well with Aqua and other applications.

4.2.2. Configuration

After downloading and compiling, a certain amount of configuration needs to be done, but the Administrator's Guide at http://emboss.sourceforge.net/docs/adminguide/ is very helpful.

Similarly, the end-user should be familiar with the documentation at http://emboss.sourceforge.net/docs/. The user will want to decide whether they are running the software as a single user or setting up for multiple users. A detailed and useful Frequently Asked Questions document (FAQ) is included with the distribution and should be read in conjunction with the online guides.

EMBOSS reads configuration data from a number of places:

- a file `emboss.default` in the share/EMBOSS subdirectory of the installation

- a file `.embossrc` in the directory specified by the EMBOSSRC environment variable

- a file `.embossrc` in the user's home directory

- a file `.embossrc` in the current directory.

The `emboss.default` file is created by copying `emboss.default.template`. Care needs to be exercised when modifying this file, which contains sequence database descriptions, but again the Guide contains enough information for the most inexperienced user[23]. Separate users can modify their EMBOSS preferences by editing their own `.embossrc` files.

If you are installing EMBOSS for multiple users, it is possible that you have sets of users who need to share specific setups. This may, for example, take the form of providing access to different sets of databases or the need to use a different data directory. It can be time consuming and error prone to maintain a series of individual EMBOSS configuration files, or to force users to work in the same directory, or to copy a `.embossrc` to each directory they wish to work in. Similarly, some databases have restrictive licences and access to them may need to be controlled.

The environment variable $EMBOSSRC can be set to point to an arbitrary directory containing an .embossrc which can then be used to give workgroup specific configuration. Each user then only needs to set EMBOSSRC in their own shell configuration (.cshrc for csh or .profile for bash) to get the workgroup specific setup.

The use of environment variables also has the advantage that only a single line of code needs to be changed if, for whatever reason, the location of the database(s) changes.

A word about SRS:

Most 'standalone' and home users will want to access remote databases, rather than having to mount and index the CD-based distributions of each sequence database. Although the Administrator's Guide does contain information on setting this up ('method:url') the syntax of the Sequence Retrieval System (SRS) changes with every release, and it is best to check <http://srs.ebi.ac.uk/linking2SRS.html> and associated pages first. The recommended procedure is to construct a query by hand in your browser, check that it returns the expected result, and then use that as a template for the DB declaration(s) in emboss.default, substituting '%s' for the search code in the manual construct.

A very useful query string of which to be aware is the EMBL 'all text' search method. This allows the user to search for a text term within the EMBL database. The appropriate declaration for emboss.default would be:

```
DB srs_alltext [
  type: N
  format: embl
  method: url
  dbalias: srs_alltext
  url: "http://srs.ebi.ac.uk/cgi-bin/wgetz?-e+[EMBL-AllText:%s]"
  comment: "text search against EMBL using EBI SRS server"
  httpversion: '1.1'
]
```

and to search for 'UBA' we would use the following command (the convention for representing a UNIX command line prompt followed by user input is `% user input` and I shall use this construction henceforth):

```
% seqret srs_alltext:UBA
```

This will write all the sequences that contain the term 'UBA' into a single (1.1 MB!) file.

Mac and Solaris users should note that using seqret to retrieve sequences from remote databases can be problematic. The maintainers are aware of the problem, and in the meantime users should copy and paste sequences from other sources (e.g. a web browser) into a text file, which seqret and the other programs can handle in the normal way.

4.2.3. Using EMBOSS

Whether running as a lone user and administrator or using a pre-installed, site-wide package, using the EMBOSS programs is the same. Those familiar with the GCG package should feel at home with the way EMBOSS operates — essentially only the names have changed. Other users who are not afraid of the command line will soon pick it up. For others, running the programs from the command line really is not difficult and should be tried at least twice.

4.2.3.1. Graphical interfaces

For those who can not or will not use a command line interface, there are a number of graphical interfaces (GUIs) available. The java-based Jemboss http://emboss.sourceforge.net/Jemboss/ is bundled with the latest versions of EMBOSS and should be installed and configured by your system administrator. Other GUIs are in development and can be found at http://emboss.sourceforge.net/interfaces/. These interfaces have the advantage that they can be used in a server/client relationship for (for example) accessing the EMBOSS programs remotely through a web browser in the absence of being able to telnet or ssh to the host machine. Furthermore, it should be possible to set up a single Mac under OS X so that it serves the EMBOSS package to machines running different (older) systems. This brings the benefits of EMBOSS to users of legacy hardware. 'There are also widgets that allow developers and users to create customized interfaces using Perl and Python.

The specifics of the setting up and use of these interfaces are beyond the scope of this discussion and are treated comprehensively elsewhere.

4.2.3.2 Running the programs

Having gained access to an installation of EMBOSS (or having installed and configured it yourself) the first thing to do is probably find out what programs are available to you. It is a good idea to work through the tutorial [24], but I will briefly describe a typical workflow here.

All EMBOSS programs are launched by typing their name at the command line. Although they run interactively, they can also be controlled with command line switches or 'qualifiers', which modify the program's behaviour or specify input and output files, etc. One of the most important command line switches is '-help', which (perhaps unsurprisingly) yields online help.

So, for example, typing `wossname -h` at a UNIX prompt lists all the possible command line switches;

```
% wossname -h
Mandatory qualifiers:
[-search] string                  Enter a word or words here and a case-
                                  independent search for it will be made in the
                                  one-line documentation of all of the EMBOSS
                                  programs. If no keyword is specified,all pro
                                  grams will be listed.

Optional qualifiers (* if not always prompted):
  -explode              boolean   The groups that EMBOSS applications belong
                                  to have two forms, exploded and notexploded.
                                  The exploded group names are more numerous
                                  and often vaguely phrased than the non-
                                  exploded ones. The exploded names are formed
                                  from definitions of the group names that
                                  start like NAME1:NAME2 and which are then
                                  expanded into many combinations of the names
                                  as: 'NAME1','NAME2','NAME1 NAME2',NAME2
                                  NAME1'. The non-expanded names are simply
                                  like: 'NAME1 NAME2'.

  -outfile              outfile   If you enter the name of a file here then
                                  this program will write the program names and
                                  brief descriptions into that file.

  -html                 boolean   If you are sending the output to a file,
```

etc.

Similarly, `% tfm programname` is the equivalent of the UNIX man ('manual') command;

```
% tfm wossname
Displays a program's help documentation manual
                             wossname

Function
  Finds programs by keywords in their one-line documentation
Description
  This allows a user to search for keywords or parts of words in the
  brief documentation (as displayed by a program when it first starts).
  The program name and the brief description is output. If no words to
  search for are specified, then details of all the EMBOSS programs are
  output.
```

etc.

As with all EMBOSS programs, running wossname with or without qualifiers brings up a brief description of that program's function, before any action is taken;

```
% wossname
Finds programs by keywords in their one-line documentation
Keyword to search for, or blank to list all programs:
```

The program wossname is probably the most useful of all EMBOSS programs. When run without qualifiers it lists all EMBOSS programs grouped by function, with a brief description of each. This serves as a handy reminder to the user of the names of programs - as the programs are called from the command line by name - , and also can help answer the question, 'What do I do now with my sequence?'.

Possibly the second most useful program is seqret[25], which we met briefly when discussing SRS in 2.2.1 above. Although seqret's one line description is very simple, this is an incredibly versatile program; it is often the first step in sequence analysis as it can take almost any sequence format[26] and convert it into something useful. Its further uses and a full description of how to use it are described in reference[25]. A full list of programs and links to their help files may be found at http://emboss.sourceforge.net/apps/.

4.2.3.3 A worked example

Rather than searching through that list, or even the programs grouped by function[6], let us return to wossname to remind us what to do next, having used seqret to read in a sequence. What we want to do is translate the sequence, but we can not remember the program used. So, we instruct wossname to return a list of all programs with 'translate' in their description:

```
% wossname translate

Finds programs by keywords in their one-line documentation
SEARCH FOR 'TRANSLATE'
backtranseq     Back translate a protein sequence
prettyseq       Output sequence with translated ranges
transeq         Translate nucleic acid sequences
```

So we can see that the program we are looking for is transeq. Taking our seqret example, we can separate the single sequence we want from the other 62 hits;

```
% seqret uba.fasta:bc004904 bc004904.seq
Reads and writes (returns) sequences
```

and write it into the file bc004904.seq (doing an SRS fetch - % seqret embl:bc004904 bc004904.seq- achieves the same thing).

Now we want to translate the sequence, starting at position 74 and ending at 1933;

```
% transeq -regions=74-1933
Translate nucleic acid sequences
Input sequence(s): bc004904.seq
Output sequence [bc004904.pep]: tap.pep
```

We could have put everything on one line; % `transeq bc004904.seq tap.pep -reg=74-1933`. Note also how the qualifier '-regions' can be abbreviated to an unambiguous three letters. The file `tap.pep` now contains our translated protein:

```
% cat tap.pep
>BC004904_1 Homo sapiens nuclear RNA export factor 1, mRNA (cDNA clone MGC:4612
IMAGE:3504065), complete cds.
MADEGKSYSEHDDERVNFPQRKKKGRGPFRWKYGEGNRRSGRGGSGIRSSRLEEDDGDVA
MSDAQDGPRVRYNPYTTRPNRRGDTWHDRDRIHVTVRRDRAPPERGGAGTSQDGTSKNWF
KITIPYGRKYDKAWLLSMIQSKCSVPFTPIEFHYENTRAQFFVEDASTASALKAVNYKIL
DRENRRISIIINSSAPPHTILNELKPEQVEQLKLIMSKRYDGSQQALDLKGLRSDPDLVA
QNIDVVLNRRSCMAATLRIIEENIPELLSLNLSNNRLYRLDDMSSIVQKAPNLKILNLSG
NELKSERELDKIKGLKLEELWLDGNSLCDTFRDQSTYISAIRERFPKLLRLDGHELPPPI
AFDVEAPTTLPPCKGSYFGTENLKSLVLHFLQQYYAIYDSGDRQGLLDAYHDGACCSLSI
PFIPQNPARSSLAEYFKDSRNVKKLKDPTLRFRLLKHTRLNVVAFLNELPKTQHDVNSFV
VDISAQTSTLLCFSVNGVFKEVDGKSRDSLRAFTRTFIAVPASNSGLCIVNDELFVRNAS
SEEIQRAFAMPAPTPSSSPVPTLSPEQQEMLQAFSTQSGMNLEWSQKCLQDNNWDYTRSA
QAFTHLKAKGEIPEVAFMK*
```

4.2.3.3.1 Options

Rather than remembering every qualifier/command line switch for each program, or having to run `program -help` each time, a useful command line switch to use is '-options'. For example, we can format the output from `wossname` into HTML, perhaps for writing a local user's guide with links to program help files;

```
% wossname -opt
Finds programs by keywords in their one-line documentation
Keyword to search for, or blank to list all programs:
Use the expanded group names [N]: y
Output program details to a file [stdout]: wossname.html
Format the output for HTML [N]: y
String to form the first half of an HTML link: localhost/
String to form the second half of an HTML link: .html
Output only the group names [N]:
Output an alphabetic list of programs [N]:
```

The file `wossname.html` contains HTML-formatted output;

```
% more wossname.html
<h2><a name="2D STRUCTURE">2D STRUCTURE</a></h2>
<table border cellpadding=4 bgcolor="#FFFFF0">
<tr><th>Program name</th><th>Description</th></tr>
<tr><td><a href="localhost/einverted.html">einverted</a></td><td>Finds DNA
inverted repeats</td></tr>
<tr><td><a href="localhost/garnier.html">garnier</a></td><td>Predicts protein
secondary structure</td><
/tr>
<tr><td><a
href="localhost/helixturnhelix.html">helixturnhelix</a></td><td>Report nucleic
acid binding
motifs</td></tr>
```

etc.

The '-options' option is also useful for reminding us what parameters we might need to consider, as in this example using transeq;

```
% transeq -opt
Translate nucleic acid sequences
Input sequence(s): bc004904.seq
Translation frames
           1  :  1
           2  :  2
           3  :  3
           F  :  Forward three frames
          -1  :  -1
          -2  :  -2
          -3  :  -3
           R  :  Reverse three frames
           6  :  All six frames
Frame(s) to translate [1]:
Genetic codes
           0  :  Standard
           1  :  Standard (with alternative initiation codons)
           2  :  Vertebrate Mitochondrial
           3  :  Yeast Mitochondrial
           4  :  Mold, Protozoan, Coelenterate Mitochondrial and
Mycoplasma/Spiroplasma
           5  :  Invertebrate Mitochondrial
           6  :  Ciliate Macronuclear and Dasycladacean
           9  :  Echinoderm Mitochondrial
          10  :  Euplotid Nuclear
          11  :  Bacterial
          12  :  Alternative Yeast Nuclear
          13  :  Ascidian Mitochondrial
          14  :  Flatworm Mitochondrial
          15  :  Blepharisma Macronuclear
          16  :  Chlorophycean Mitochondrial
          21  :  Trematode Mitochondrial
          22  :  Scenedesmus obliquus
          23  :  Thraustochytrium Mitochondrial
Code to use [0]:
Regions to translate (eg: 4-57,78-94) [1-2264]: 74-1933
Trim trailing X's and *'s [N]: y
Change all *'s to X's [N]:
Output sequence [bc004904.pep]: tap.pep
```

4.2.3.4 What now?

EMBOSS is a complete sequence analysis package, and what you want to do with your sequence — multiple sequence analysis, pattern recognition, etc. — is up to you. The tutorial[24] has further worked examples and suggestions for productive workflow. One should also be aware of the EMBASSY [3, 27] program suite, which extends the EMBOSS core package. Joe Felsenstein's popular phylogenetic package *PHYLIP*[28, 29] is an important part of EMBASSY, which also includes *emnu* (a character-based menu for display and selection of EMBOSS programs), *mse* (a multiple sequence editor) and *topo* (for displaying transmembrane maps of proteins).

4.3 How do I get involved?

The first thing to do is to download and install the software if appropriate, and to start using it! EMBOSS is still under development, and programs are still being improved and added to it. Just by using the software you might find bugs and 'features' that no one else has picked up on. While I was preparing this chapter I discovered a set of circumstances that caused pepcoil to crash. Within an hour of submitting a bug report to <emboss-bug@embnet.org> I had a work around and a promise that the problem would be fixed in the next release.

End-users should sign up to the surprisingly low-traffic *<emboss@embnet.org>* open mailing list (see http://emboss.sourceforge.net/support/#usermail), for announcements and discussion of the package. There are searchable archives at http://www.rfcgr.mrc.ac.uk/Emboss/HYPERMAIL/emboss/.

The EMBOSS team actively solicits contributions from developers. Existing programs that could be integrated into EMBOSS are required (and free beer is promised for authors!) as are developers who will code new programs and extend/improve existing functions. There is a closed mailing list for developers[30], and instructions for getting involved[31] and accessing the CVS tree[32]. New interfaces[14] to the package are also solicited.

The one downside to all this is that the UK Medical Research Council has announced the closure of the Research and Bioinformatics divisions of the Rosalind Franklin institute. This does not mean that EMBOSS will cease to exist in the middle of 2005 but it does raise doubts about its continued active development. The programs will (probably) remain available for download but the British Government will not be funding further work on the package. However, the embracing of Open Source ideology at the inception of the EMBOSS project has probably ensured its continued survival as there is no licensing barrier to prevent voluntary developers maintaining and improving the package. The Staden package has already been released as Open Source[33], and the possibility of integrating Staden's sequence and trace data manipulation tools with EMBOSS is real. Further work on EMBOSS will depend on the continued involvement of developers throughout Europe and elsewhere. This is a matter of high priority for scientists, as EMBOSS remains a suitable, integrated replacement for GCG and a host of individual programs, and is a valuable resource for end-users in the biology lab.

Acknowledgements

Thanks to David Martin and Peter Rice for helpful discussion and bug-hunting.

References

1. http://www.embnet.org/
2. http://www.hgmp.mrc.ac.uk/Software/EMBOSS/Apps/index.html
3. EMBASSY, http://www.hgmp.mrc.ac.uk/Software/EMBOSS/EMBASSY/
4. http://www.hgmp.mrc.ac.uk/Software/EMBOSS/Themes/SequenceFormats.html#in
5. http://www.hgmp.mrc.ac.uk/Software/EMBOSS/Themes/SequenceFormats.html#out
6. http://www.hgmp.mrc.ac.uk/Software/EMBOSS/Apps/groups.html
7. http://www.sanger.ac.uk/
8. http://www.rfcgr.mrc.ac.uk/
9. http://www.accelrys.com/
10. http://www.pgpi.org/
11. http://emboss.sourceforge.net/what/
12. http://www.compbio.ox.ac.uk/CBRG_home.shtml
13. http://emboss.sourceforge.net/docs/
14. http://emboss.sourceforge.net/interfaces/
15. http://staden.sourceforge.net/
16. http://www.gnu.org/copyleft/lesser.html
17. UNIX is a registered trademark of The Open Group <http://http://www.opengroup.org/> in the United States and other countries.
18. http://www.faqs.org/docs/jargon/U/Unix.html
19. http://www.rfcgr.mrc.ac.uk/About/closure.html, http://www.freeshell.org/
20. http://www.apple.com/macosx/
21. http://www.apple.com/scitech/stories/osxporting/
22. http://fink.sourceforge.net/
23. http://www.hgmp.mrc.ac.uk/Software/EMBOSS/Doc/Admin_guide/adminguide/node4.html
24. http://emboss.sourceforge.net/docs/emboss_tutorial/
25. http://www.hgmp.mrc.ac.uk/Software/EMBOSS/Apps/seqret.html
26. http://www.hgmp.mrc.ac.uk/Software/EMBOSS/Themes/SequenceFormats.html
27. http://www.hgmp.mrc.ac.uk/Software/EMBOSS/Apps/index.html#embassy
28. http://www.hgmp.mrc.ac.uk/Software/EMBOSS/EMBASSY/PHYLIP/index.html
29. http://evolution.genetics.washington.edu/phylip.html
30. emboss-dev@embnet.org, http://www.hgmp.mrc.ac.uk/Emboss/HYPERMAIL/emboss-dev/
31. http://emboss.sourceforge.net/developers/
32. http://emboss.sourceforge.net/developers/cvs.html
33. http://www.mrc-lmb.cam.ac.uk/pubseq/news.2003.html

Chapter 5

Gene Family Bioinformatics: a Study Case with Annexin A7

Reginald O. Morgan and Maria-Pilar Fernandez

Contact: Reginald O. Morgan, Ph.D.
Reginald O. Morgan and Maria-Pilar Fernandez
Department of Biochemistry and Molecular Biology, Edificio Santiago Gascon,
Faculty of Medicine, University of Oviedo, E-33006 Oviedo, Spain
E-mail: romorgan@bioquimica.uniovi.es

5.1 Introduction

The gene family is a central focus for research into the molecular basis of life. Knowledge of its structural and evolutionary profile can reveal significant insight into phenotypic and functional relationships between species or biological processes. Homologous genes with a common ancestry encode proteins with structural similarities and related functions that are particularly amenable to systematic and computational analyses. The scope, detail and significance of those relationships can be clarified through comparative sequence analysis and visualization tools for emerging genomic data, even where curated annotations and biological data may be lacking. The term function is applicable not only to the interaction of the protein end products encoded by DNA, but to all the regulatory molecules involved in the timing, level and localization of gene expression patterns. The primary goal of bioinformatics has been to identify, annotate, classify and compare new genomes, within which protein encoding gene families represent a miniscule material fraction but a major functional contribution to all living organisms. The methodology described here is intended to demonstrate how bioinformatics can be selectively applied to gene family studies as an integral component of empirical research with the ultimate goal of elucidating (patho)physiological function.

Studies of molecular evolution were originally pioneered by mathematical biologists with interest in species phylogeny and classification, but the information content of the gene family as a molecular marker for tracing species and chromosomal history has now generated a broad gamut of sophisticated software programs to investigate the structure, function and evolution of the gene family itself. Applied bioinformatics has become a vital research tool that extends beyond the mere management of voluminous and complex genomic data by attempting to interpret them, decipher mechanisms, generate and test hypotheses. A fundamental concept of evolution is that diversifying changes which contribute to the survival fitness of an organism will ultimately become fixed into the population. Functional constraint or purifying selection has thus guided the pattern and process of gene family evolution, as a "viability filter" for otherwise stochastic gene duplication events and spontaneous mutations. As a consequence, the molecular records being catalogued by genome sequencing projects contain recognizable information on structural elements that are conserved across species and inferred to be critical for function and survival. Conservation and divergence in molecular evolution can be calibrated against geological or species fossil records to estimate event rates and times, and the contribution of structural change to functional divergence can be measured even at single sites within DNA or proteins from related subfamilies. A dual challenge for gene family bioinformatics is therefore to identify key functional elements in both regulatory and protein coding regions based on their conservation level in species orthologs, and also to discern which differences among paralogous genes are responsible for functional divergence in expression regulation or protein interaction.

As an exemplary model, the bioinformatic analysis of the annexin A7 (ANXA7 gene) subfamily can serve to demonstrate a practical approach to molecular systematics and computational analysis of gene families and the value of such information toward understanding gene function. Although the gene was discovered 25 years ago[1] and much is known about the structural evolution of other family members[2] there is still limited understanding of the structure-function relationships and regulatory role of annexins[3]. The initial obstacle is to identify as many homologues as is feasible to refine the structural profile for more extensive search, classification and comparison. Rigorous phylogenetic analysis and other computational approaches are important to perceive structural and functional relationships and to estimate the time-scale for key molecular events such as gene duplication. Chromosomal linkage maps and the characterization of gene features and organization in regulatory and coding regions can provide important corroborative evidence for phylogenetic analysis to trace historical relationships of genes of different species, i.e. to differentiate orthologous equivalents in different species or paralogous gene duplications within a single species. Gene expression analysis is a particularly important aspect of functional genomics and a formidable challenge for bioinformatics in the form of microarray clustering, statistical analysis of expressed sequence tags, and the *in silico* prediction of key promoter regulatory elements prior to direct empirical testing. Finally, *in silico* modeling techniques are increasingly effective for visualizing and evaluating key molecular interactions between proteins, protein-DNA, and chromatin.

5.2 Molecular Profiling

A basic ingredient for the sequence-based bioinformatic analysis of gene families is a replete set of properly aligned sequences representing true homologs. Annotated sequences for certain reference species such as human, mouse and rat can generally be retrieved from centralized databases such as GENECARDS (see website URLs below for bioinformatics resources) and pairwise sequence searches for close homologs with BLAST[4] and FASTA program families are feasible at major bioinformatics centers such as the National Center for Biotechnology Information (NCBI) and the European Bioinformatics Institute (EBI). More extensive species searching may require varied techniques such as protein searching of nucleotide databases, cyclical "PSI-BLAST" for ancestral sequences, or the use of domain profiles to capture distant isoforms within a particular family. However, the bulk of gene family data presently remains as unidentified genomic or expressed sequence tag (EST) DNA sequences in databases derived from hundreds of ongoing genome sequencing projects, for which data download and local searching with installed programs is most effective. Whole genome shotgun "trace" sequences, high throughput genome sequences (HTGS), genome survey sequences (GSS) and equivalent resources are generally available via http or ftp internet protocols at publicly accessible sites.

Using the aforementioned public resources, the number of annexin A7 sequences available for the present study was extended from 4 (human, mouse, rat and frog sequences in GenBank/EMBL/DBJ) to 14 full-coding cDNA sequences, and from 2 partial (human and mouse) to 8 complete gene sequences spanning a broad range of vertebrate and nonvertebrate animal species. The most useful format for systematic and computational analyses is the multiple sequence alignment (MSA), assembled by standard programs such as CLUSTALW[5] and visualized with basic phylogenetic analysis programs such as GENEDOC, SEAVIEW or PHYLOWIN[6]. Homology (i.e. common ancestry) of clustered sequences is an important concern in gene family studies and can be addressed by appropriate comparative analysis of complete and accurate MSA. Pairwise sequence shuffling algorithms (see FASTA) are helpful, bootstrap analysis or clustering by neighbor joining of MSA (CLUSTALW) is generally more effective, and full statistical resolution can be achieved by rigorous phylogenetic analysis (see later). However, one of the most effective new approaches for statistically characterizing a MSA is the construction of a molecular profile based on algorithms incorporating hidden Markov models[7]. The nucleotide (nt) or amino acid (aa) MSA of annexin A7 coding core region sequences for 14 taxa were processed by HMMER to yield file models describing site-specific statistics for these homologous cDNA and protein segments. The information contained in the protein core HMM model can be visualized as a sequence logo (Figure 5-1) as produced by LOGOMAT[8] that summarizes conservation level (by full stack height), aa frequency (by respective symbol height) and majority consensus (top aa) for each site in the sequence. The logo facilitates interpretation of sites where selective evolutionary constraint has preserved structure (hence function) and the HMM model (i.e. alignment matrix) also has practical applications in searching sequence databases and extending MSA or developing profile libraries for entire gene families or individual subfamilies to use as a statistically reliable gene classification tool. One important application of this probabilistic methodology has been the elaboration of annotated databases such as PFAM[9] or SUPERFAMILY, which define structural domains by HMM profiles for thousands of known protein families.

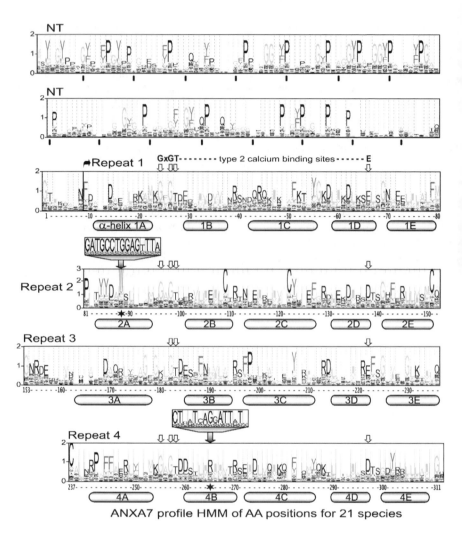

ANXA7 profile HMM of AA positions for 21 species

Figure 5-1. Sequence logo for a hidden Markov model of the annexin A7 subfamily. The complete coding sequences for 14 vertebrate orthologs of annexin A7 were compiled from public databases, aligned by CLUSTALW, processed by HMMER, and converted to a sequence logo for the homologous tetrad core region by LOGO-MAT. The full stack height at each site reflects overall conservation level and inferred functional importance while the contribution of individual aa is shown in relative scale with the majority consensus sequence on top. A similar nt logo is shown for 2 sites of special significance for the presence of a highly conserved motif containing a unique Trp88 and an apparent single nt polymorphism (SNP) encoding an Arg265Gln conversion in a residue considered crucial for calcium channel ion selectivity.

The HMM model of the core region in ANXA7 proteins (depicted as a logo in Figure 5-1) clearly distinguishes the 4 internally aligned repeats of the homologous tetrad and the elevated conservation of the type 2 motif for calcium coordinating residues in each repeat. A comparison with HMM models and logos constructed for all 12 subfamilies in vertebrate "A" family annexins identified unique sites such as Trp88 in a highly conserved nt domain, and sequence searching of the various models has proven to be a powerful instrument for the preliminary identification of candidate orthologs from earlier diverging, nonvertebrate species (see later). A second nucleotide inset in Figure 5-1 at position Arg265 identifies a single nucleotide polymorphism (SNP) of potential importance, based on the conservation level and localization of this residue in the physical model of this putative calcium channel protein. The universally conserved Arg in this and other annexins is encoded by the AGG codon in most annexin A7 genes from amphibian to rhesus monkey (AGA in chicken) but has apparently converted to CGG in chimpanzee and human. Some genomic sequences for human ANXA7 have discovered a G/A polymorphism (e.g. see rs3750575 in dbSNP at NCBI) at the second base of the codon for this locus, resulting in a CAG codon encoding glutamine, which represents a significant property change in the key residue considered to be the main voltage sensor for annexin calcium channels. Since SNPs are common in the genome (ca. every 1000 bp) and frequently clustered, their locations in coding or regulatory regions should be further evaluated in the context of possible functional consequences and population statistics.

5.3 Phylogenetic Analysis

An overview of the relationships between orthologous genes from different species, and their broader relationship with other paralogous genes duplicated within the entire (super)family, can best be achieved through phylogenetic analysis. Excellent program packages such as PHYLIP, PAUP, PAML and TREE-PUZZLE utilize the MSA to trace site-specific conservation for tree-building by maximum likelihood, parsimony, distance matrix, bootstrapping and various other algorithms (see PHYLIP site for information and links). A MSA of the coding tetrad for annexin A7 from 14 vertebrate species and related outgroup taxa was analyzed by TREE-PUZZLE (maximum likelihood resolved by quartet puzzling) using either *aa*, the first 2 bases of each codon or all nt to obtain a partially resolved phylogenetic tree (Figure 5-2). The branching order corresponds to the known species separation times because earlier diverging species retain relatively more ancestral characters despite their relatively longer branch lengths, which reflect amount of evolution. The 14 species of ANXA7 defined by the HMM model are fully resolved among primates, placental and marsupial mammals, bird and amphibian, in which the pseudotetraploid *Xenopus* harbors a lineage-specific duplication after its separation from *Silurana* frogs. Due to saturated mutation of the third codon base allowed by coding redundancy, the use of aa or the first 2 nt of the codon is recommended to reduce "noise" in tracing site-specific changes throughout the taxa, especially for distant relatives. Program parameter selection is important for obtaining statistically valid results (i.e. ML or bootstrap values at nodes) on which to base phylogenetic inferences, and the inherent need to allow for site-specific variation should be evident from the results of Figure 5-1.

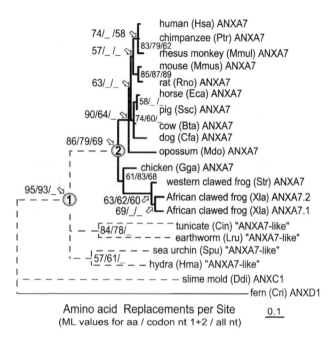

Figure 5-2. Phylogenetic tree for the annexin A7 subfamily. Maximum likelihood (ML) analysis was performed by TREE-PUZZLE with a gamma distribution for site-specific rate variation in 311 aa (topology shown), codon nt 1+2, or all nt of the homologous tetrad core region. All bifurcations were resolved with ML values shown for 14 confirmed vertebrate orthologs of annexin A7. Significant basal separation was also established for related invertebrate homologs, designated only as "ANXA7-like", and for outgroups from slime mold ANXC1 and plant fern ANXD1 at the root.

We have also identified 4 potential orthologs of ANXA7 in earlier diverging species of urochordate, annelid worm, echinoderm and Cnidaria, for which the basal divergence position in phylogenetic analysis (Figure 5-2), sequence comparison with the HMM model (Figure 5-1) and molecular dating (see later) are consistent with this assignment. However, we cautiously designate these only as "ANXA7-like" at this time until certain discrepancies are resolved to provide physical corroboration, such divergent amino termini in *Hydra* and sea urchin, the annexin A11-like gene structure of the Ciona annexin, and the unavailability of genetic linkage data for any of these species. Interestingly, the slime mold annexin ANXC1, originally considered to be a putative ortholog of annexin A7[10], is clearly in a separate clade from ANXA7 and more closely associated with the plant outgroup, fern annexin D1. This phylogenetic clade separation is now corroborated by evidence that gene structures and genetic linkage maps are entirely distinct for ANXC1 and ANXA7 (see later), despite any similarity in their (nonhomologous) amino termini or more remote common ancestry.

5.4 Computation of Molecular Dates and Functional Constraints

A surprising observation from the MSA and its phylogenetic analysis is the apparent absence of ANXA7 in fish! It is poignant to note that most of the 12 vertebrate annexins in "A" family" do have representatives in up to 16 species of fish now being sequenced, each confirmed by dozens of ESTs, hundreds of genomic trace sequences, and generally as two gene copies resulting from the ancient tetraploidization event detected early in the lineage of bony fishes. Moreover, the lack of any evidence from sequence searches using the unique amino termini and coding exon sizes (see later) specific to ANXA7 strongly support the conclusion that annexin A7 really does not exist in fish, and this negative finding is very near certainty for the nearly complete genomes of zebrafish and 2 pufferfish. This is an important question to resolve because data from two *Anxa7* gene knockout experiments in mice suggest that gene loss is either homozygous lethal[11] or associated with viable alterations of cellular calcium metabolism[12]. Even if fish do not have the same need for annexin A7 and earlier diverging relatives from invertebrate animals cannot yet be confirmed to be true orthologs, the question arises as to when annexin A7 originated as a distinct gene from other annexins. Did the gene not exist prior to the divergence of fish approximately 450 million years ago (Mya) or was it selectively "lost" from teleosts as a consequence of genome reorganization? Partial or global phylogenetic analyses with up to 300 orthologous and paralogous members of "A" family annexins have already established that the annexin A7 branch bifurcates with annexin A11 from a common ancestor with annexin A13 as the most basal annexin in a rooted tree, and that the fish annexins branch much later in each of the annexin A11 and annexin A13 gene lineages[13 and unpublished]. Molecular dates can be extrapolated for the nodes from which these individual gene lines emerge in a phylogenetic tree (computed with or without a molecular clock to normalize branches) by quantifying branch lengths and calibrating them against branch lengths for orthologous species with separation times dated by the fossil record.

An alternative computational approach to molecular dating is to estimate the number of nonsynonymous nucleotide substitutions (NNS, expressed as *aa* changes) between orthologous sequences as a measure of protein divergence and calibrate the separation time of the host species based on fossil records to obtain an average gene evolution rate. The divergence time between two paralogous genes is then given by the average number of NNS between the two genes in different species divided by twice the average substitution rate for the two genes. NNS estimates, corrected for multiple hits by programs such as K-ESTIMATOR[14] were computed for annexin A7 to quantify the increasing extents of its divergence between rodents and other mammals, between birds and mammals, and for amphibians from mammals (leftmost panels of Figure 5-3). The ratio of NNS causing aa changes to silent synonymous nucleotide substitutions (SNS) yields a measure of the functional constraint on the former, such that there are 10-20 times fewer NNS than SNS in orthologous sequence comparisons of annexin A7 (middle panels of Figure 5-3). That is, purifying selection has effectively filtered out more than 90% of the nt substitutions in annexin A7 that would have caused aa changes detrimental to the survival fitness of the host organisms, a level which testifies to the importance of conserving annexin protein structures to preserve function[15]. The NNS divergences between the above-mentioned species groups, coupled with generally accepted separation times of 110, 310 and 360 million years (Myr) for other mammals, birds and amphibians from humans provide 3 independent estimates for the evolution rate of the annexin A7 coding tetrad (ca. 0.155 NNS/site/1000 Myr in the rightmost panels of Figure 5-3). Finally, NNS estimates of the divergence between ANXA7, ANXA11 and ANXA13 together with calculated rates of evolution for each gene lead to the conclusion that annexins A7 and A11 were probable duplication products of annexin A13 approximately 700 Mya [13, 15]. This is consistent with the existence of (putative) orthologs of annexin A7 in the invertebrate species identified in Figure 5-2 and leads to the conclusion that ANXA7 probably existed prior to the emergence of the fish lineage, which must therefore have suffered selective loss of this gene.

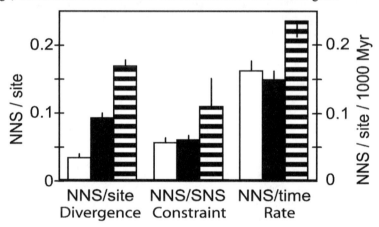

Figure 5-3. Nucleotide substitution computation of protein divergence, selective constraint and gene evolution rates for the annexin A7 subfamily. Nonsynonymous (NNS) and synonymous (SNS) nt substitutions were estimated with standard deviation for the homologous tetrad core regions of anxa7 cDNA by comparison of rodent sequences with other mammals (open blocks, n=8), birds-mammals (solid, n=6), and amphibians-mammals (striped, n=18) using K-ESTIMATOR. The average NNS measures the extent of protein core evolution between the respective species groups (leftmost 3 columns), the NNS/SNS ratio measures expressed versus silent changes reflecting selective functional constraint on protein change (middle), and orthologous annexin A7 gene rates were normalized by dividing the NNS divergences between the 3 species groups by their generally accepted separation times from human lineage of 110, 310 and 460 Myr for rodents, birds and amphibians, respectively.

5.5 Genetic Maps

The chromosomal localization of genes is fundamental to understanding genetic history, function and pathophysiological role. Most human and mouse annexins were mapped cytologically and the genetic linkage maps from these models have been used to assemble genomic sequence and provide physical maps for these and other emerging genomes. Human *ANXA7* in chromosome 10q21.1[16] and mouse *Anxa7* in chromosome 14[17] are contained within extensive chromosomal homology groups and are syntenic with annexins A8 and A11 on the respective chromosomes in both species. Genomic BLAST searches with either cDNA or gDNA sequences on the NCBI site now establish physical locations for these genes (Figure 5-4) but fail to confirm the existence of a pseudogene purported to reside in mouse chromosome 10 [17]. Rat *Anxa7* was similarly localized near the P-terminus of chromosome 15 in a much smaller linkage group and attended by confirmed pseudogenes on chromosomes 2, 5, 7 and X. Differences in the order and orientation of mammalian linkage groups are consistent with intrachromosomal rearrangements affecting annexin A7, possibly including a common ancestral origin with annexin A11. Since genetic linkage can be useful for corroborating or refuting orthology, we also verified that there was no correspondence between *ANXA7* genes and *Dictyostelium anxc1* in chromosome 1 or *anxc2* in chromosome 4 and that flanking mammalian genes like protein phosphatase 3 (*PPP3CB*) are remotely located in *Dictyostelium* chromosome 2. Any functional relationship between annexin A7 and its neighboring genes remains to be established. The evolution from classical to molecular genetics has been accompanied by increased interest in population variation to account for allelic variants that might dispose to altered phenotypic or pathological states. There is growing interest in the systematic and computational bioinformatic analysis of population data for single nucleotide polymorphisms (dbSNP), such as the coding SNP rs3750575 discussed above, and chromosomal aberrations in cancer (MITELMAN) with relevance to annexin A7[18].

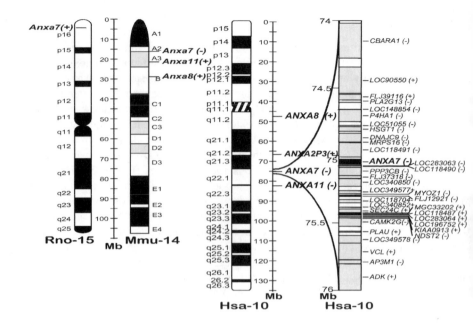

Figure 5-4. *Genetic maps (ideogram and physical) for annexin A7 loci in human chromosome 10, mouse-14 and rat-15. Data were obtained from NCBI MAPVIEWER by BLAST searches of each genome (builds 34, 32 and 2 respectively) with cDNA and gene sequences. Loci are contained in extensive, homologous linkage groups, with a partial gene list displayed (rightmost) for bp 74-76 Mb of human chromosome 10. There were no cDNA matches to possible pseudogenes in the human or mouse genomes (cf. literature), but at least 4 Anxa7 rat pseudogenes were confirmed by sequence analysis of segments on rat chromosomes 2, 5, 7 and X (not shown).*

5.6 Gene Organization

Comparative genomics can be useful for assessing structural divergence and its temporal appearance to elucidate both genome evolution and gene function. The gene structures characterized for human *ANXA7*[16] and mouse *Anxa7*[17] determined a unique exon splicing compared to annexin A13 and congeners of annexin A11[13]. Using the fourteen characterized cDNAs for annexin A7 (above), we searched and compiled genomic trace data to assemble complete contiguous gene structures for annexin A7 in 7 species and confirm the same splice patterns in remaining species (Figure 5-5). This established the evolutionary stability of the annexin A7 gene structure since amphibian separation 360 Mya and the utility of its exon splicing pattern as a unique evolutionary marker to search for the gene in other species. It corroborated the conclusion that ANXA7 is absent from fish but was partially successful in identifying the first invertebrate annexins with a phylogenetically close relationship to ANXA7 (Figure 5-2) and "ANXA7-like" gene structures (Figure 5-5).

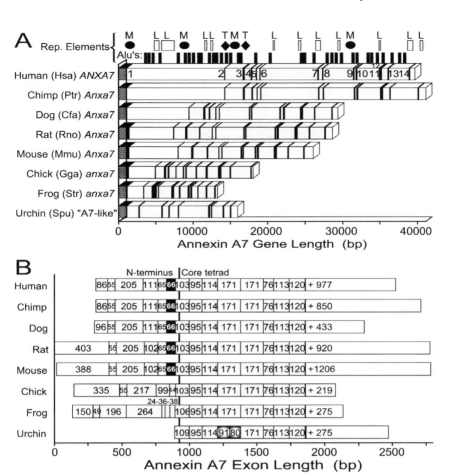

Figure 5-5. *Comparative structural genomics of annexin A7 genes from 8 species — repetitive elements, exon distribution and exon sizes. Full-length cDNAs were reconstructed from BLAST searches of dbEST and used to identify and extend exons in whole genome shotgun trace sequences, which were concatenated into full-length gene consensus sequences using DNASIS, SEQTOOL or contig analysis utilities. A. Complete gene sequences were first screened by REPEATMASKER to map internal repetitive elements, shown on top for human ANXA7 with symbols for 38 Alu SINEs, 13 LINE1, L2 or L3 elements (L), 2 MaLR Long Terminal Repeats (T), and 4 MER1 DNA elements (M) located in the 5'upstream promoter region. The total gene size and distribution of numbered exon bands are drawn to scale. B. Exon sizes are aligned to illustrate congruency in the coding (shaded) core regions and similarity in amino terminal regions. The transcription start point(s) in exon 1 and multiple polyA sites for 3' termination are based on the most extensive matching cDNAs from dbEST or UniGene.*

A comparative analysis distinguishes the relatively small exons 1 and variable 3' polyadenylation sites in non-rodent mammals, with concordance between empirical data and *in silico* analysis. The presence of an alternatively spliced 66 bp exon 6 is seen only in mammals, and an apparently truncated amino terminus and extra intron splice characterize the "A7-like" sea urchin annexin. The extensive variation in exon size and amino acid composition encoded by the 5' terminus diminishes the reliability of this region as a molecular marker for this gene in earlier diverging species, such as sea squirts, sea urchins, *Hydra* or *Dictyostelium*, for which N-terminal similarities and composition are an insufficient basis to establish orthology.

Annexin A7 gene size has apparently increased in more recently diverged species (w.r.t. human) due to intron expansion and the incorporation of genomic repetitive elements that are largely species specific. The 38 Alu SINEs, 13 LINE elements, 2 Long Terminal Repeats and 4 MER DNA elements comprise 53% of the 40 kb human ANXA7 gene with a 41% average GC content and show only limited correspondence to orthologous genes from other mammals. The 26 kb mouse *Anxa7* gene, for example, contains only 30% repetitive elements and 43% average GC content, including 35 "B" SINE elements, 3 LINEs, 4 LTRs and 3 MER elements distinct from those found in human. One of the best, informative resources for comparative genomic viewing of bioinformatically deduced structures and maps of gene family members is the genome browser at UCSC[19]. Detailed studies of many species genomes may, however, be required to clarify how underlying intronic mutations or species differences in spliceosome machinery account for alternative exon splicing in certain annexin A7 genes, and whether the insertion of repetitive elements has played a significant role in species-specific changes of *ANXA7* gene structure or regulation. The specificity in expression and function of the coding cassette exon in brain, heart and skeletal muscle[20], observed only in mammals, can be further interpreted from the bioinformatic analysis of gene expression microarray data (see following).

5.7 Expression Profiling

Extensive gene expression profiles are now available from dbEST and microarray studies through portals such as SOURCE[21] and the NCBI Gene Expression Omnibus. These provide useful perspective on which to base hypotheses and experiments for the study of mechanisms in gene expression regulation. As a starting point we have examined public microarray data for human and mouse annexin A7 in normal tissues, supplied by LSBM (Reference Database for Gene Expression Analysis), READ (Riken Expression Array Database)[22] and GNF (Gene Expression Atlas)[23] (URLs listed below). These indicate that annexin A7 is a moderately expressed gene in relation to other annexins and that it exhibits no marked tissue specificity (Figure 5-6). The modest expression level and ubiquity in both species suggest functional equivalence but this contrasts with gross structural disparity of the promoter regions in these species affecting sequence alignment, localization of transcription start points and the identification of regulatory elements *in silico* and *in vitro* (see later). The apparent lack of tissue-specific variation and poor inducibility of annexin A7 under different cell conditions emphasize the potential utility of microarray data and the need for cross-platform bioinformatic comparisons to implicate changes in annexin A7 gene expression with a particular cellular state or regulatory pathway.

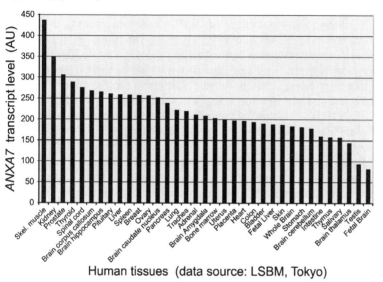

Human tissues (data source: LSBM, Tokyo)

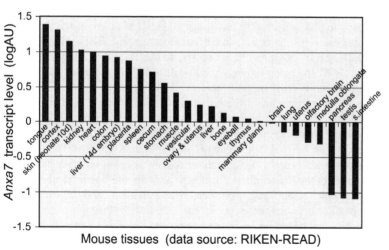

Mouse tissues (data source: RIKEN-READ)

Figure 5-6. Annexin A7 expression transcript levels (arbitrary units, AU) in normal human and mouse tissues screened by matrix microarray. Human data based on GeneChip® HG-U133A technology were retrieved from the LSBM (Tokyo) Reference Database for Gene Expression Analysis and mouse data from the READ - Riken Expression Array Database[22]. Signal values for selected normal tissues are drawn on the same arbitrary scale. Corroborating data are available from the GNF site.

The absence of tissue-specific gene expression is at odds with the relative abundance of the larger 51 kD protein isoform produced by alternative exon splicing in muscle and brain[20]. One of the myriad of specialized EST and microarray studies of tissue-specific gene expression profiles in cancer has confirmed the exclusive expression of the higher mass isoform of ANXA7 in cancer irrespective of tissue[24] suggesting that the mechanism probably involves variations in the cellular spliceosome machinery rather than a defined pathophysiological role. The decreased ANXA7 expression in severe stages of certain cancers has further suggested that it may act as tumor suppressor, although broader changes in the expression profile of other annexins and different genes now suggest that the change is more likely to be a consequence rather than an etiological factor[18, 25]. Evidence for post-transcriptional regulation of annexin gene expression[26] suggests that comparisons of transcript and protein expression microarrays might also be fruitful in identifying models for study of annexin mRNA turnover rates. The vital role for bioinformatics in microarray analysis will come to focus more on the statistical validation and construction of interacting gene networks to decipher the role of enigmatic genes like the annexins.

5.8 Characterization of Promoter Features

The structural characterization of annexin A7 genes (Figure 5-5) and assessment of global expression profiles (Figure 5-6) focus inquiry about the molecular architecture and functionality of the 5'upstream promoter region (Figure 5-7). General features were analyzed for sequence similarity, nt composition and C+G/A+T ratios using basic bioinformatic tools, CpG islands were characterized by CpGProD[27], insertion of genomic repetitive elements by REPEATMASKER, transcription start site(s) assignments aided by dbTSS[28] and nt polymorphisms documented in dbSNP. Interspecies comparisons identified some conserved features relevant to function and other divergent features that could explain discrepancies. A recent functional characterization of the human *ANXA7* promoter[29] found that low expression level was due in part to repressor activity in the immediate 5'flanking nontranscribed region and there was some cellular dependency in this repressive effect. The presence of MER DNA elements in the human upstream promoter, a canine SINE in dog and several B4 elements in rodents account for poor homology between rodent and other mammalian promoters and for the apparent absence of similarity with the chicken and frog promoters. These elements and the variation in the CpG island profiles between these species may account for the significant shift in transcription start points observed *in vitro* and predicted *in silico* and may question the suitability of mouse as a model for annexin A7 gene regulation. The bioinformatic characterization of promoter structural elements is a crucial prelude to the design of empirical studies of gene expression and knockout models.

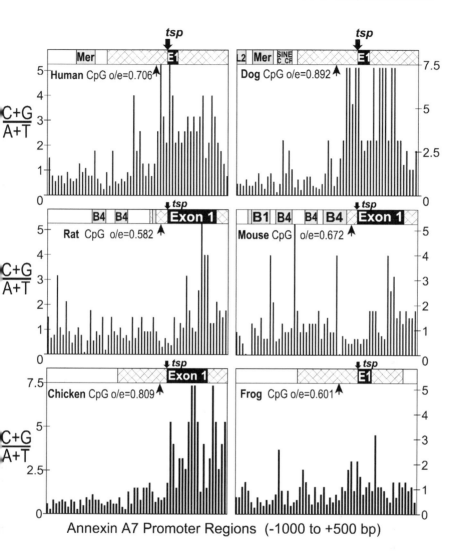

Annexin A7 Promoter Regions (-1000 to +500 bp)

Figure 5-7. Structural features of promoter regions for annexin A7 genes from verte-brate species. The designation of transcription start points was based on public expression data encoded by exon 1 in dbEST or from DBTSS[28] for human and mouse, and 1000 bp upstream plus 500 bp downstream (i.e. exon 1 and 5'intron 1) were mapped. Repetitive elements were identified by REPEATMASKER (Mer5A, SINE C_CF, B1 and B4 in top bars), CpG island statistics were computed by CpGProD [27], and the C+G/A+T ratio histogram was calculated for 25 bp windows. Note the pres-ence of distinctive repetitive elements, variable sizes for exon 1 and divergent CpG island profiles in the various species.

5.9 Phylogenetic Footprinting of Regulatory Elements

Sequence similarity and alignment may not be a valid basis for inferring common functionality of orthologous promoters because of architectural rearrangements resulting, for example, from the introduction of species-specific repetitive elements (Figure 5-7). The bioinformatic evaluation of putative regulatory elements has become much more sophisticated with matrix matching rather than identity matching of binding sites characterized by HMMs for transcription factors. Genomatix MATINSPECTOR identified at least 18 predefined regulatory elements common to annexin A7 promoters, including two 7-element models with marked interspecies conservation of Sp1 (expected for CpG island promoters), MyoD and MEF2 (relevant for expression in muscle[30], ZBPF and ZFIA (both zinc finger core promoter element) and EGRF binding sites. CONSITE attempts to identify potential transcription factor binding sites in orthologous promoters by defining conserved regions in an aligned sequence pair, delimited by known cDNA, that contain threshold match scores to matrices for known regulatory elements, such as the c-Rel element immediately upstream of the tsp in several annexin A7 promoters. The concept of regulatory "modules" that define cooperative sets of elements interacting in a spatial or temporal genomic context represents a significant further advance in defining promoter function, especially where suitable complex models have been characterized in detail experimentally. However, when there is limited evidence *a priori* to implicate specific regulatory elements, a new bioinformatic approach termed phylogenetic footprinting (by analogy to the experimental approach) may be particularly useful. It attempts to identify suspect elements on the basis of their conservation and frequency in orthologous promoters from a range of species resulting from the functional constraint on evolutionary change. Since there is little apparent homology for annexin A7 promoters between rodents, other mammals, birds and amphibians, it is especially important that any search method not be strictly based on physical location or extensive sequence alignment.

The annexin A7 promoters from eight species were analyzed by FOOTPRINTER [31] to identify and map conserved motifs common to all species, thereby describing a set of candidate regulatory elements independent of gross sequence similarity, location and the limited database of known transcription factors (Figure 5-8). These included the 6 species described in Figure 5-7 plus chimpanzee and rhesus macaque monkey promoters, which were 99% and 91% identical to human, respectively. As illustrated in the sequence logo output for dog (Figure 5-8B) some conserved blocks were found to lie within known genomic repetitive elements in the upstream region or coinciding with the splice site for intron 1, but others with varying degrees of cross-species conservation were found in the proximal upstream promoter and especially the downstream transcribed region. The elements highlighted by font size and shade were common to all annexin A7 promoters in the 8 species examined. Although the GC-rich nature of the CpG islands in annexin A7 promoters (Figure 5-7) contribute some key elements, such as Sp1 binding sites around the *tsp*, the relative importance of downstream elements in exon 1 and 5'intron 1 of all species had not been previously recognized. The results can thus facilitate planning and interpretation of experimental promoter studies, including nuclease protection assays to identify true transcription factors relevant to annexin A7 expression in different species. The CONSITE webserver attempts to achieve similar results by pairwise promoter sequence comparisons.

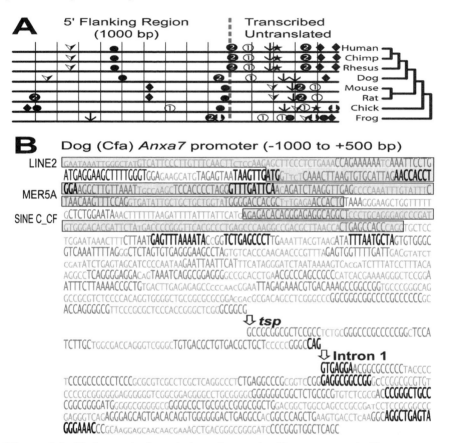

Figure 5-8. *Phylogenetic footprinting of annexin A7 promoters. A. The promoter regions (-1000 to +500 bp relative to the putative tsp) of the annexin A7 gene from human, chimpanzee, rhesus monkey, dog, mouse, rat, chicken and Silurana frog were analyzed together with the corresponding species tree by FOOTPRINTER[31]. Potential regulatory motifs were identified on the basis of cross-species sequence conservation and equivalent sites were physically mapped using matching symbols. B. Sample output for the dog Anxa7 promoter is shown with motif height and intensity reflecting the parsimony conservation score for conserved elements, some potentially acting as putative transcription factor binding sites. Repetitive genomic elements identified by REPEATMASKER and specific to dog are gray-shaded in the 5' distal sequence.*

5.10 Molecular Modeling of Evolutionary Information

The structural and functional divergence of paralogous gene families is due partly to major chromosomal events such as duplication or repetitive element insertion and partly to diffuse changes resulting from the spontaneous nt mutation rate. However, regulatory control of gene expression and protein interactions ultimately determine, through evolutionary selection, which changes are compatible with function and viability. The conservation level of specific aa sites in orthologous proteins of a gene subfamily can thus be inferred to reflect the functional importance of those sites. Significant changes in the rate of evolution at other sites detected by subfamily comparisons may instead be indicative of structural and functional divergence between gene members, independent of speciation. Bioinformatic tools that identify and quantify either conservation or change from the MSA and phylogenetic trees of gene families can provide novel insight into the functionally important regions of a particular subfamily and the mechanisms for its divergence. CONSURF[32] and RATE4SITE[33] were used to perform maximum parsimony and maximum likelihood analysis of MSA for 14 orthologous ANXA7 protein sequences together with information from the phylogenetic tree to calculate indices of evolutionary conservation for each homologous site. DIVERGE[34] performed similar analysis to detect sites affected by evolutionary rate change distinct from other annexin subfamilies. Such evolutionary information can be mapped as a color or shading scheme onto the residues of a model derived from crystallographic analysis or sequence threading of close structural homolog. Many excellent protein modeling programs are available for this purpose, including SWISS-MODEL and DEEPVIEW [35], RasMol and RasTop, MolMol, GRASP and others.

The HMM consensus sequence for ANXA7 from 14 vertebrate species with 87-91% aa identity (Fig. 1) was threaded through the crystal structure coordinates of its closest structural homolog, *Hydra* annexin 12[36] with 54% aa identity (Figure 5-9). Many of the conserved sites identified by CONSURF and RATE4SITE coincided with those observed for other annexins, although some of special interest for ANXA7 include the juxtaposition of an unusual Trp88 with the Arg265Gln SNP and the colocalization of Cys109 and Cys151, especially well-conserved in ANXA7 (viz. black-shaded atoms in Figure 5-9). Those residues in ANXA7 exhibiting significant evolutionary changes in rate and/or physicochemical property during speciation, detected by RATE4SITE and DIVERGE, are distinguished from other annexin subfamilies by the space-filled atoms in varying shades of gray. Sites distinct from both close relatives annexin A11 and A13 include Ser82,Thr83, Gln164, Val33 and Leu216 (dark gray, numbered as in Fig. 1) Sites diverged from the founding member annexin A13 were mainly charged residues among Arg121, Asp130, Gln165, Arg172, Ala204, Arg209, Arg221, Lys232, Ala283, Val310, while divergence from its closest homolog annexin A11 is more confined to uncharged residues (lighter gray) Leu12, Ile17, Ala49, Glu118, Arg136, Glu205, Ala208, Leu239, Gln280 and Met285. In summary, the structural conservation of annexin A7 evident in NNS calculations (above) can be determined on a site-specific basis and visualized in 3-dimensional protein models to understand intrinsic function, and patterns of rate divergence from other family members can be useful for interpreting the pleiotropic properties of each subfamily.

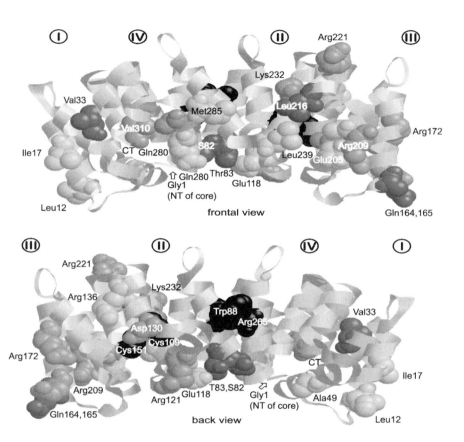

Figure 5-9. *Protein surface mapping of evolutionary determinants. The majority consensus sequence derived from an HMM model representing 14 vertebrate ANXA7 proteins (see Figure 5-2) was threaded through the crystal structure coordinates of Hydra annexin 12* [1AEI-PDB, 36] *by the SWISS-MODEL*[35] *webserver and visualized as a ribbon model with RASMOL. The space-filling residues were shaded by a RASMOL script based on information defining site-specific evolutionary conservation by RATE4SITE*[33] *and evolutionary rate changes by DIVERGE*[34]*. The lower figure is laterally inverted and numbers correspond to sites of the core tetrad (Fig. 1). NT denotes N-terminus. Atoms in solid black are uniquely conserved in ANXA7, those in dark and medium gray correspond to the degree of site-specific divergence from ANXA11 and ANXA13 proteins.*

5.11 Conclusion

The comparative analysis of the annexin A7 subfamily here represents an exemplary demonstration of how applied bioinformatics can relate evolutionary conservation to function and evolutionary change to divergence in speciation or pathology. Gene family researchers can make a significant primary contribution to current genome annotation efforts by developing gene (sub)family profiles based on HMMs and confirming relationships by phylogenetic analyses as a framework for nomenclature schemes. Other computational analyses can be applied to determine the timing and rates of key events such as gene duplication, repetitive element insertion or pseudogene frequency. Systematic comparisons of genomic and proteomic data from diverse databases can elucidate map locations, gene structures and expression profiles to observe gene function in a broader context. The pattern and process of annexin A7 evolution in a range of species has revealed the constancy of certain molecular markers such as gene structure and genetic linkage that contrast with more variable features such as amino terminal homology, alternative exon splicing, site-specific divergence in the tetrad core region and possibly in clinically relevant SNPs. Attention to both site-specific sequence conservation and change can be informative and cross-disciplinary studies can accelerate the elucidation of gene regulation, protein function and evolutionary history.

Phylogenomics (the integration of evolutionary biology with molecular genomics) has amplified the need and utility for bioinformatics to transform information into knowledge. Molecular evolutionary analysis provides a rational basis for understanding gene relationships and for correlating structural changes with the functional diversification of gene families. Such bioinformatic analysis, also termed interpretative genomics, requires thorough and efficient identification of all available homologs along with their patterns and rates of change to formulate inferences about the evolutionary and biological processes that underlie them. Computer hardware and software and the bioinformatic algorithms that move them have become vital, integral components of modern biomedical research because information management may now overshadow data content for advancing science. Isolated findings can be viewed with multidisciplinary perspective by scientists adept at utilizing informatics tools and by informaticians interested in biological systems, with an eye to deciphering complex spatial and temporal relationships involved in the coming era of systems or cyberbiology.

References

1. Creutz, C.E., Pazoles, C.J., Pollard, H.B.: Identification and purification of an adrenal medullary protein (synexin) that causes calcium-dependent aggregation of isolated chromaffin granules. *J Biol Chem* 1978;253:2858-2866.
2. Fernandez, M.P., Morgan, R.O.: Structure, function and evolution of the annexin gene superfamily; in Bandorowicz-Pikula J (ed): The Annexins - Biological Importance and Annexin-related Pathologies. *Landes Bioscience*, 2003, pp 21-37.
3. Gerke, V., Moss, S.E.: Annexins: from structure to function. *Physiol Rev* 2002;82:331-371.
4. Altschul, S.F., Madden, T.L., Schaffer, A.A., Zhang, J., Zhang, Z., Miller, W., Lipman, D.J.: Gapped BLAST and PSI-BLAST: A new generation of protein database search programs. *Nucl Acids Res* 1997;25:3389-3402.
5. Thompson, J.D., Higgins, D.G., Gibson, T.J.: CLUSTAL W: improving the sensitivity of progressive multiple sequence alignment through sequence weighting, positions-specific gap penalties and weight matrix choice. *Nucleic Acids Res* 1994;22:4673-4680.
6. Galtier, N., Gouy, M., Gautier, C.: (1996) SEAVIEW and PHYLO_WIN: two graphic tools for sequence alignment and molecular phylogeny. *Comput Appl Biosci* 1996;12:543-548.
7. Eddy. S.R.: Profile hidden Markov models. *Bioinformatics* 1998;14:755–763.
8. Schuster-Böckler, B., Schultz, J., Rahmann, S.: HMM logos for visualization of protein families. BMC Bioinformatics 2004, 5:7 (e-pub).
9. Bateman, A., Birney, E., Cerruti, L., Durbin, R., Etwiller, L., Eddy, S.R., Griffiths-Jones, S., Howe, K.L., Marshall, M., Sonnhammer, E.L.: The Pfam protein families database. *Nucl Acids Res* 2002;30:276-280.
10. Greenwood, M., Tsang, A.: Sequence and expression of annexin VII of Dictyostelium discoideum. *Biochim Biophys Acta* 1991;1088:429-432.
11. Srivastava, M., Atwater, I., Glasman, M., Leighton, X., Goping, G., Caohuy, H., Miller, G., Pichel, J., Westphal, H., Mears, D., Rojas, E., Pollard, H.B.: Defects in inositol 1,4,5-trisphosphate receptor expression, Ca(2+) signaling, and insulin secretion in the anx7(+/-) knockout mouse. *Proc Natl Acad Sci USA*. 1999;96:13783-13788.
12. Herr, C., Smyth, N., Ullrich, S., Yun, F., Sasse, P., Hescheler, J., Fleischmann, B., Lasek, K., Brixius, K., Schwinger, R.H., Fassler, R., Schroder, R., Noegel, A.A.: Loss of annexin A7 leads to alterations in frequency-induced shortening of isolated murine cardiomyocytes. *Mol Cell Biol* 2001;21:4119-4128.
13. Iglesias, J.M., Morgan, R.O., Jenkins, N.A., Copeland, N.G., Gilbert, D.J., Fernandez, M.P.: Comparative genetics and evolution of annexin A13 as the founder gene of vertebrate annexins. *Mol Biol Evol* 2002;19:608-618.
14. Comeron, J.M.: K-Estimator: Calculation of the number of nucleotide substitutions per site and the confidence intervals. *Bioinformatics* 1999;15:763-764.
15. Morgan, R.O., Bell, D.W., Testa, J.R., Fernandez, M.P.: Genomic locations of ANX11 and ANX13 and the evolutionary genetics of human annexins. *Genomics*

1998;48:100-110.
16. Shirvan, A., Srivastava, M., Wang, M.G., Cultraro, C., Magendzo, K., McBride, O.W., Pollard, H.B., Burns, A.L.: Divergent structure of the human synexin (annexin VII) gene and assignment to chromosome 10. *Biochemistry* 1994;33:6888-6901.
17. Zhang-Keck, Z.Y., Srivastava, M., Kozak, C.A., Caohuy, H., Shirvan, A., Burns, A.L., Pollard, H.B.: Genomic organization and chromosomal localization of the mouse synexin gene. *Biochem. J.* 1994;301:835-845.
18. Srivastava, M., Montagna, C., Leighton, X., Glasman, M., Naga, S., Eidelman, O., Ried, T., Pollard, H.B.: Haploinsufficiency of Anx7 tumor suppressor gene and consequent genomic instability promotes tumorigenesis in the Anx7(+/-) mouse. *Proc. Natl. Acad. Sci. USA* 2003;100:14287-14292.
19. Kent, W.J., Sugnet, C.W., Furey, T.S., Roskin, K.M., Pringle, T.H., Zahler, A.M., Haussler, D.: The human genome browser at UCSC. *Genome Res.* 2002;12:996-1006.
20. Magendzo, K., Shirvan, A., Cultraro, C., Srivastava, M., Pollard, H.B., Burns, A.L.: Alternative splicing of human synexin mRNA in brain, cardiac, and skeletal muscle alters the unique N-terminal domain. *J. Biol. Chem.* 1991;266:3228-3232.
21. Diehn, M., Sherlock, G., Binkley, G., Jin, H., Matese, J.C., Hernandez-Boussard, T., Rees, C.A., Cherry, J.M., Botstein, D., Brown, P.O., Alizadeh, A.A.: SOURCE: a unified genomic resource of functional annotations, ontologies, and gene expression data. *Nucl. Acids Res.* 2003;31:219-223.
22. Bono, H., Kasukawa, T., Hayashizaki, Y., Okazaki, Y.: READ - RIKEN Expression Array Database. *Nucleic Acids Res.* 2002;30:211-213.
23. Su, A.I., Cooke, M.P., Ching, K.A., Hakak, Y., Walker, J.R., Wiltshire, T., Orth, A.P., Vega, R.G., Sapinoso, L.M., Moqrich, A., Patapoutian, A., Hampton, G.M., Schultz, P.G., Hogenesch, J.B,: Large-scale analysis of the human and mouse transcriptomes. *Proc. Natl. Acad. Sci. USA* 2002;99:4465-4470.
24. Xu, Q., Lee, C.: Discovery of novel splice forms and functional analysis of cancer-specific alternative splicing in human expressed sequences. *Nucleic Acids Res.* 2003;31:5635-5643.
25. Smitherman, A.B., Mohler, J.L., Maygarden, S.J., Ornstein, D.K.: Expression of annexin I, II and VII proteins in androgen stimulated and recurrent prostate cancer. *J. Urol.* 2004;171:916-920.
26. Raynal, P., Pollard, H.B., Srivastava, M.: Cell cycle and post-transcriptional regulation of annexin expression in IMR-90 human fibroblasts. *Biochem. J.* 1997;322:365-371.
27. Ponger, L., Mouchiroud, D.: CpGProD: identifying CpG islands associated with transcription start sites in large genomic mammalian sequences. *Bioinformatics* 2001;18:631-633.
28. Suzuki, Y., Yamashita, R., Nakai, K., Sugano, S.: DBTSS: DataBase of human Transcriptional Start Sites and full-length cDNAs. *Nucl. Acids Res.* 2002;30:328-331.
29. Srivastava, M., Pollard, H.B.: Low in vivo levels of human anx7 (annexin VII)

gene expression are due to endogenous inhibitory promoter sequences. *Cell Biol. Int.* 2000;24:475-481.

30. Clemen, C.S., Hofmann A, Zamparelli C, Noegel AA: Expression and localisation of annexin VII (synexin) isoforms in differentiating myoblasts. J *Muscle Res. Cell Motil.*: 1999;20:669-679.
31. Blanchette M, Tompa M. Discovery of regulatory elements by a computational method for phylogenetic footprinting. *Genome Res.* 2002;12:739-748.
32. Glaser, F., Pupko, T., Paz, I., Bell, R.E., Bechor, D., Martz, E., Ben-Tal, N.: ConSurf: Identification of functional regions in proteins by surface-mapping of phylogenetic information. *Bioinformatics* 2003;19:163-164
33. Pupko, T., Bell, R.E., Mayrose, I., Glaser, F., Ben-Tal, N.: Rate4Site: an algorithmic tool for the identification of functional regions in proteins by surface mapping of evolutionary determinants within their homologues. *Bioinformatics* 2002;18:S1-S7.
34. Gu, X.: Functional divergence in protein (family) sequence evolution. *Genetica* 2003;118:133-141.
35. Schwede, T., Kopp, J., Guex, N., Peitsch, M.C.: SWISS-MODEL: an automated protein homology-modeling server. *Nucleic Acids Res.* 2003;31:3381-3385.
36. Luecke, H., Chang, B.T., Mailliard, W.S., Schlaepfer, D.D., Haigler, H.T.: Crystal structure of the annexin XII hexamer and implications for bilayer insertion. *Nature* 1995;378:512-515.

Website URLs Relevant to Gen(om)e Bioinformatics

BCM: Baylor College of Medicine sequencing center and sequence utilities: [http://searchlauncher.bcm.tmc.edu/].

CONSITE: Phylogenetic footprinting of orthologous promoter sequence pairs, Center for Genomics and Bioinformatics, Karolinska Institute, Sweden: [http://www.phylofoot.org/].

CONSURF: Phylogenetic mapping of protein site conservation: [http://consurf.tau.ac.il/].

CpGProD (CpG Island Promoter Detection), [http://pbil.univ-lyon1.fr/software/cpg-prod.html].

DBTSS (Database of Transcriptional Start Sites), [http://dbtss.hgc.jp/].

DEEPVIEW Swiss PDB Viewer (authors Guex N, Diemand A, Peitsch MC, Schwede T): [http://www.expasy.ch].

DIVERGE (author Gu X, computation of site-specific evolutionary rate changes to measure protein functional divergence) [http://xgu1.zool.iastate.edu].

DOE-JGI (Department of Energy Joint Genome Institute) sequencing center: [http://www.jgi.doe.gov/].

EBI (European Bioinformatics Institute) bioinformatics center: [http://ebi.ac.uk].

FOOTPRINTER (phylogenetic footprinting) [http://bio.cs.washington.edu/software.html].

GENECARDS (gene encyclopedia) [http://genecards.weizmann.ac.il].

GENEDOC (alignment viewer): [http://www.psc.edu/biomed/genedoc].
GENOMATIX (program suite for promoter analysis), [http://www.genomatix.de/].
GNF (Genome Novartis Foundation), [http://expression.gnf.org/cgi-bin/index.cgi].
Goldenpath comparative genomic mapping: [http://genome.ucsc.edu].
HGNC (HUGO Gene Nomenclature Committee) [http://www.gene.ucl.ac.uk/nomenclature/]
HMM (hidden Markov models) software: [http://hmmer.wustl.edu/].
Stanford Genomic Resources, [http://genome-www.stanford.edu/].
Keck Center for Comparative and Functional Genomics, University of Illinois at Urbana-Champaign, [http://www.biotech.uiuc.edu/keck.htm].
LOGOMAT (HMM sequence logo webserver, Max Planck Institute for Molecular Genetics) [http://logos.molgen.mpg.de/].
LONGHORN (Stanford microarray database) [http://www.longhornarraydatabase.org/index.html].
LSBM (Laboratory for Systems Biology and Medicine, Univ. Tokyo, Microarray Reference Database for Gene Expression Analysis: [http://www.lsbm.org/index_e.html].
MATINSPECTOR (promoter analysis for transcription factor binding sites), [http://genomatix.gsf.de/products/index_mat.html].
MIRAGE (gene expression regulation): [http://www.ifti.org/].
MITELMAN (Database of Chromosomal Aberrations in Cancer): [http://cgap.nci.nih.gov/Chromosomes/Mitelman].
NCBI (National Center for Biotechnology Information): [http://ncbi.nlm.nih.gov].
NCI60 (National Cancer Institute microarray): [http://genome-www.stanford.edu/nci60/index.shtml].
PAML (Phylogenetic Analysis by Maximum Likelihood) [http://abacus.gene.ucl.ac.uk/software/paml.html].
PAUP (Parsimony Analysis), [http://paup.csit.fsu.edu/].
PBIL protein bioinformatics (Pôle Bio-Informatique Lyonnais): [http://pbil.univ-lyon1.fr].
PFAM (Protein family HMM database) [http://pfam.wustl.edu/].
PHYLIP (Phylogeny Inference Package, distributed by the author, Felsenstein J, since 1980, Department of Genetics, University of Washington, Seattle: [http://evolution.genetics.washington.edu/phylip.html].
R8S software (phylogenetic tree analysis for divergence rates and times): [http://ginger.ucdavis.edu/r8s/].
RASMOL (Bernstein HJ, molecular modeling graphics) [http://www.bernstein-plus-sons.com/software/RasMol_2.7.1].
RASTOP (Valadon P, protein modeling) [http://www.geneinfinity.org/rastop/].
RATE4SITE (surface mapping of evolutionary information into 3D protein structures): [http//ashtoret/tau.ac.il].
READ (Riken Expression Array Database), [http://read.gsc.riken.go.jp/].

REPEATMASKER (Smit A, Green P: unpublished, scanner for repetitive elements): [http://ftp.genome.washington.edu/cgi-bin/RepeatMasker].

SEQTOOLS (author: Søren W Rasmussen) [http://www.dnatools.org/seqtools.htm].

SANGER (Institute, genome sequencing center: [http://www.sanger.ac.uk/].

SOURCE [http://source.stanford.edu/cgi-bin/sourceSearch].

SUPERFAMILY gene classification [http://superfam.org].

SWISS-MODEL [http://www.expasy.org/swissmod].

UCSC Genome Browser (University of California Santa Cruz, Genome Bioinformatics) [http://genome.ucsc.edu/].

WEIZMANN Institute Bioinformatics, [http://bioinformatics.weizmann.ac.il/].

WICGR (Whitehead Institute Center for Genome Research) sequencing center: [http://www.wi.mit.edu/].

WUSTL-GSC (Washington University at St. Louis Genome Sequencing Center) sequencing center: [http://genome.wustl.edu/gsc/].

Chapter 6

The Analysis of DNA and Protein Sequences Using Desktop Software

Robert H. Gross, Ph.D.

Contact: Robert Gross, Ph.D.

Associate Professor, Department of Biological Sciences and Director, Center for Biological and Biomedical Computing, Gilman Labs 104, Dartmouth College, Hanover, New Hampshire 03755 USA
E-mail: Robert.H.Gross@Dartmouth.edu
E-mail: bob@textco.com

6.1 Introduction

DNA and protein sequence analyses used to be the purview of mainframe computers. The casual biologist did not often have access to terminals through which they could interact with the analysis code, and even if they did have access, getting programs to do what the user wanted was often very complex. Times have changed. There are now a wealth of relatively easy to use sequence analysis packages available to run on desktop or laptop computers. The results of these analyses are the same as if they were run by an expert on a mainframe computer many years ago. This change has been brought about by the increase in computing power available on the desktop and by the desire of program developers to create software that can be understood and used by the average laboratory biologist.

Along with the opportunity to perform sophisticated analyses on the desktop, comes the responsibility to interpret intelligently the output of the analyses. This requires some basic understanding of the algorithms behind the analyses and careful analysis of the results. Many molecular biology courses include lectures on this topic these days. The combination of this understanding and the availability of desktop analysis software has truly opened up numerous opportunities for researchers to explore sequence information and glean biological insights.

Current desktop sequence analysis software is easier to use than mainframe programs that often have a command line driven interface or have a data input that is the simplest interface necessary to satisfy the needs of multiple computer platforms and operating systems that might be used as clients. With today's powerful laptop computers it is also possible to take your sequence analysis projects on the road. Having worked on mainframes and stared at green letters on a black background, I find it particularly satisfying to be able to fly cross-country at 35,000 feet and work on a complex sequence analysis project on my trusty laptop. Another advantage of current desktop sequence analysis software is that the application contains many different kinds of analyses, all presented to the user through a common interface. Once familiar with the interface, getting work done is relatively straightforward. There is no longer a need to struggle trying to figure out how to run the program but instead, one can focus on interpreting the results and thinking of biological implications.

Of course, there are still important analyses that can only be done on larger shared computer systems or clusters of UNIX boxes that run many parallel processes for multiple users simultaneously. Searching large databases or doing genomic scale comparisons and calculations are tasks that cannot be conveniently handled on desktop systems (yet) because the computer power is just not available at this time. Computer clusters with a suite of programs are often the best solution for a department of researchers all having comparable computationally intense needs. In this way, the institution can purchase the software once, maintain it centrally, and make it available to all. For individual laboratories, though, desktop sequence analysis software is usually the best solution.

This chapter focuses on desktop sequence analysis. There are a number of packages available that will do the basic analyses and all packages share some features. The applications differ in their interfaces, flexibilities, feature sets, and output styles. Some of the applications also have unique features that might be essential to the individual needs of the researcher. Perhaps the most important feature of the software is that it be intuitive to use for the biologist. If you cannot figure out how to get something done, the software is not of any use to you. Some of the better known packages are: Gene Inspector from Textco BioSoftware[1], Vector NTI from Informax[2], LaserGene from DNAStar[3], MacVector from Accelrys[4], and DNASIS from Miraibio[5].

In this chapter I will focus on Gene Inspector from Textco BioSoftware, Inc. It is available for both Macintosh OS X (an older version is available for OS 9) and Windows. Gene Inspector is representative of desktop sequence analysis software but has a number of unique features designed from the perspective of a biologist.

6.2 Gene Inspector

Designed from the beginning to be a program that is useful and intuitive for researchers, Gene Inspector is based around the concept of a molecular biologist's notebook. The notebook works as a normal laboratory notebook would work, allowing typing of comments and observations as well as displaying analysis/experiment results. The notebook metaphor allows the scientist to perform analyses and then comment on the results in adjacent text. This is similar to the process of running any normal laboratory procedure such as gel electrophoresis, pasting in an image of the gel, and then writing comments about the experiment.

In addition to the Notebook, the other kind of document that Gene Inspector creates and uses is the sequence document. DNA or protein sequences can be stored individually, or as sets of sequences that might all be part of an ongoing analytical process. By allowing multiple sequences to be stored in a single file, Gene Inspector eliminates the need to hunt down and find multiple sequence files in order to start an analysis. One or more sequences in a sequence file can be selected and used to begin with an analytical task.

All analyses are initiated through an Analysis Setup Panel that provides the same interface for setting up and running any of the more than 60 analyses included in the program. The Analysis Setup Panel provides a common route to start all analytical tasks and presents a familiar environment to the user no matter what analysis is involved.

The Notebook, multisequence files, and standard Analysis Setup Panels make Gene Inspector very simple to use because the user is never presented with an environment that is not familiar. This, in turn, enables him/her to focus on the analysis rather than on how to run the program.

Gene Inspector can import sequences from common file format such as GenBank, EMBL, GCG, and FASTA, so no matter how sequence information is obtained, it can be analyzed in Gene Inspector.

Gene Inspector allows the biologist to follow natural thought processes as well. For example, if you perform a certain analysis that yields intriguing results which suggests further investigation, you can often launch the next analytical task directly from the output results of the previous analytical routine. By allowing the researcher to follow the natural flow of ideas and questions, Gene Inspector facilitates discovery of new information. All of the analysis results will be stored in a notebook which can also contain observations of the flow of ideas that lead to the analytical tasks and results. The rest of the chapter will examine some of these concepts in more detail.

6.3 Gene Inspector Sequence Window

A sequence window contains one or more sequence(s) and can contain either DNA sequences or protein sequences (but not both). Figure 6-1 shows a DNA sequence window. Each sequence window corresponds to a specific file, in this case a file containing rhodopsin sequences. There are 4 "panes" in this window. Pane A shows an overview of the sequences that are part of this file. The length of each arrow corresponds to the length of the sequence it represents in the actual sequence listing shown in pane D. Names are shown in pane B and starting positions are shown in pane C. Each of the panes can be resized by dragging the border between the panes. It is also possible to show or hide any of the panes using menu items.

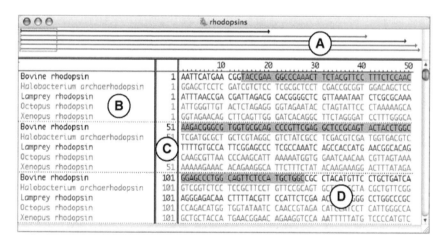

Figure 6-1.

The overview pane (A) can be used to navigate within large sequence files. Clicking any location in the overview pane will automatically scroll the display to show that location in the sequence pane (D) and will highlight the selected sequence. One can also select an entire sequence by clicking on the sequence in the name pane

(B). A selected sequence (or range of nucleotides) can be inverted (reverse complement) in which case the direction of the arrow in the overview pane is reversed.

Each sequence in the window can be edited individually. Dragging the mouse to select a sequence will skip over other sequences and will be constrained to keep the selection within the same sequence. As shown in Figure 6-1, the selection started by inserting the mouse at position 14 in the Bovine rhodopsin and dragging downwards until position 127 was reached. The program selected only the bovine sequence and skipped over the other sequences. Even though multiple sequences are present in the file, it is still a simple matter to select a range of nucleotides in only one sequence.

Other ways of selection are also feasible by using modifier keys. As shown in Figure 6-2, using the Option key (MacIntosh) or the Alt key (Windows) while dragging the mouse enables the selection of any rectangle of sequences. Command-dragging — using the Command key (MacIntosh) or Control key (Windows) — will allow the selection of a range of nucleotides in all the sequences simultaneously. These types of selections are often useful when examining multiple sequence alignments and wanting to illustrate specific regions of interest. The order of the sequences can easily be rearranged by dragging with the Option key (MacIntosh) or Alt Key (Windows) down on the name of the sequence to move it up or down to the position you desire.

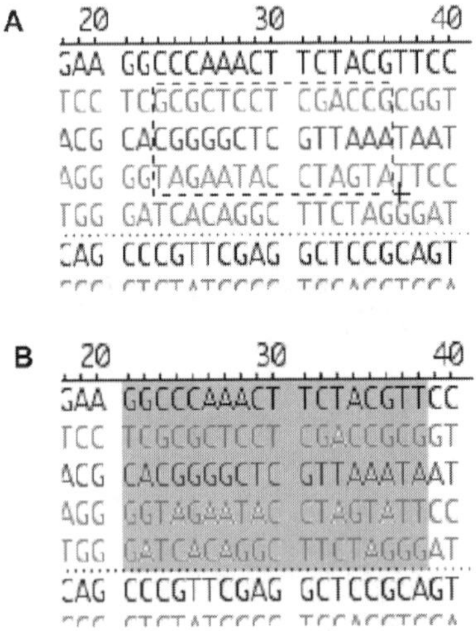

Figure 6-2.

The sequence window can contain any collection of sequences of the same type (DNA or protein). This might be a group of sequences you happen to be working with in a particular project, or it could be a collection of evolutionary related sequences, such as rhodopsins from different species. Having a group of related sequences in the same file makes it easy to perform such operations as multiple sequence alignments, which can be conducted right in the sequence window itself. Figure 6-3 shows such a multiple sequence alignment performed with a BLOSUM 62 table and comparing several lactate dehydrogenase proteins. Many tables are available for use in alignments, or custom scoring tables can be created by the user. Notice that the sequences are displayed in Times font. You are able to choose any font in your system and Gene Inspector will display all the alignments properly (you are not limited to monospaced fonts such as Courier or Monaco).

In Figure 6-3, the top row of the sequence window shows the "score" of each column as a histogram indicating the number of characters in that column that match the consensus sequence, shown in the second row. Other sequences in the window are "adorned" as defined in the Match Adornment window. The choice in this case is to invert the colors of those nucleotides that do not match the consensus sequence. This shows up as a black background for all the letters not matching the consensus. If you wanted to emphasize those characters matching the consensus, it would be better to click the other radio button and Invert characters that match the consensus. Another interesting variation is the first one, "Grade background color of characters that match the consensus sequence with []," for example, red. This option would shade the background of each column with a shade of red (the chosen color). Those columns that match the consensus completely would show a 100% red background. Those columns in which only 50% of the characters match the consensus would have a background of 50% red-50% white (in other words, a lighter shade of red). Columns with only 25% of the characters matching the consensus would be shaded with 25% red-75% white. Using the Match Adornments dialog it is possible to highlight any feature of interest in multiple sequence alignments.

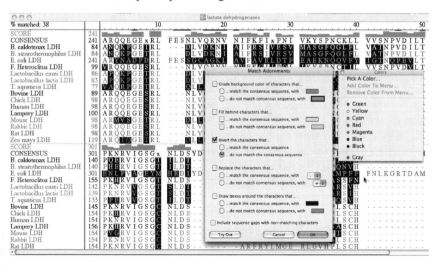

Figure 6-3.

The sequence window stores and displays multiple sequences and is also the place from which specific analyses can be initiated on sequences or segments of sequences. It has been designed to perform these functions efficiently and easily. However, there are other operations that might be desirable to perform on sequences, such as showing translated regions of DNA or indicating introns and exons. Although these operations cannot be performed in a sequence window directly, they can be done by creating a sequence object in the Notebook.

6.4 Gene Inspector Notebook

The Notebook is the other main type of Gene Inspector document. It is a word processor capable of performing the tasks of a true laboratory notebook. In addition, the Gene Inspector Notebook has some special features that distinguish it from regular word processors. Figure 6-4 shows some of the features of a Gene Inspector Notebook page. With the exception of multiple sequence alignments, every analysis conducted by Gene Inspector creates an output object that is placed in the Gene Inspector Notebook. Analysis output objects for hydropathy analysis, finding repeats, examining pH/pI behavior of a protein, helical wheel and amino acid composition are all shown (labeled with "B") in Figure 6-4. Each of these analyses shows some property of the protein being examined. Any notebook object can be assigned a Frame (border) as shown in the figure. A frame can have one, two, or three lines separated by any number of pixels. Each frame line can be assigned a color and thickness, and the frame can even have a shadow as shown in the top left two objects in Figure 6-4. Once a frame style is defined, it can be saved and then applied to any other Notebook object by selecting it from a Frames menu.

Sequence objects, labeled "A" in Figure 6-4, contain sequences either copied or dragged from a sequence window into the Notebook window. A sequence object can contain a number of features not available in the sequence window. In the bottom right sequence object, the DNA sequence has an intron (inverted color) and is also translated across that intron to show the corresponding amino acid sequence. Another possible feature in the sequence object is to mark and indicate restriction enzyme sites along the DNA (not shown). The sequence object shown along the left edge of the page is a section of a multiple sequence alignment shown with shading to indicate the extent of match.

The notebook can also contain sidebar text objects ("C" in Figure 6-4). This is a text box into which you can type. The text in a sidebar text object is separate from the background text so that it will remain on the page at the position it is originally placed unless it is manually moved. The background text will flow around objects (if that is how you want it to behave) and will expand to fill new pages as more type is entered, just like what happens in a regular word processor. In this particular Notebook page, the sidebar text object is used to display a title for the Notebook. The sidebar can also have a frame as shown in Figure 6-4.

The Gene Inspector Notebook also has a number of simple drawing tools to enable the creation of simple but useful tools objects in the notebook. The item labeled "D" in Figure 6-4 is a line object drawn directly into the Notebook. In this case, the line has an arrowhead at one end and serves to connect a comment in the background text to a particular peak in one of the analyses. At the tip of the line (arrowhead) is another Notebook tools object, an ellipse. This was used to indicate a particular peak on the output object. The judicious use of Notebook tools objects helps tremendously in clarifying discussions and indicating particular features of output objects.

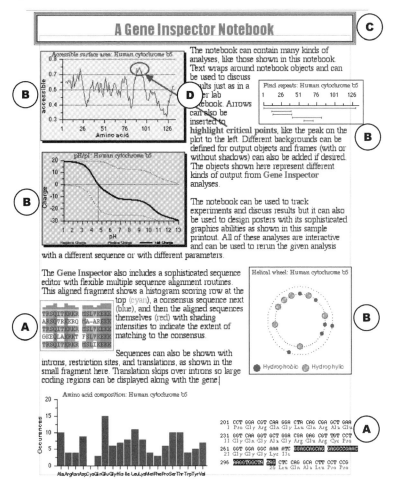

Figure 6-4.

As shown in Figure 6-5, objects from other programs can also be pasted into a Gene Inspector Notebook. In this case, a scanned image of a gel was pasted into the notebook and the legend was generated using a sidebar text object. After the gel and the sidebar were arranged, they were grouped and a rectangle (another Notebook tools object) was drawn as a border.

1. EcoR I
2. Pst I
3. Hind III
4. undigested pBG123
5. BamH I
6. Ava I
7. 1 kb ladder
8. Avr II
9. undigested pBG217
10. undigested pBG224

sequencing error (compression?) around position 3000 in the sequence. Ask Kevin to go back and check the sequencing gels at this location.

Restriction mapping of the candidate clones is shown in Figure 2.

Figure 2: Restriction digests

Figure 6-5.

Any object in a Gene Inspector Notebook can be converted into an appendix object. An appendix object may (or may not) be visible in the actual notebook. Figure 6-6 shows two appendix objects in the bottom right corner of the Notebook window. When either of these is double-clicked it opens into a separate window. This is a convenient way to store large amounts of information (e.g. a long DNA sequence) without taking up valuable Notebook space and interfering with the flow of discussion.

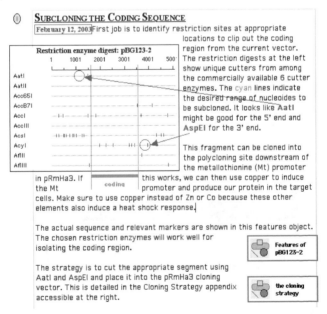

Figure 6-6.

Appendix objects can be viewed in another way, through the Notebook > Appendices menu. All appendix objects for the entire notebook are listed in this menu. Selecting an appendix from this menu will open that appendix in a new window, no matter where the cursor happens to be located in the Notebook. This is a very convenient way to store information such as the series of steps needed to perform an assay. If an assay protocol is stored as an appendix object, it can be viewed from anyplace in the Notebook, whenever the reader wants to see the protocol. This alleviates the need to repeat entries for protocols that are commonly used. The protocol is entered once as an appendix and is then always available.

Another useful feature of the Notebook is the ability to change the size of the Notebook "sheet." The Notebook sheet size is independent of the printer page size. Printer page size is defined by the printer that is selected as the currently active printer for the computer that is being used. The Notebook sheet size is the size of the surface that is being used by the Notebook for layout. This allows the user to define a sheet size of 16" x 20", for example, to use as panels for a poster presentation. When the computer is later connected to a larger format printer, the 16 x 20 pages will be printed in their entirety. Figure 6-7 shows the dialog box that allows the user to define the Notebook sheet layout. Some standard layouts are shown in the first section, allowing choices of standard text layout (8.5" x 11"), side by side layout, which has two standard sized pages adjacent to each other, and poster sheet layout. In this case, Poster sheet layout is chosen and the dimensions are set in the "Poster layout size" section as 20" wide by 16" tall. In the Columns section, it is possible to define the number of text columns to be used for the text flow. The upper right corner of the dialog shows what the actual layout looks like. Even after background text is already present in the Notebook, one can change the Notebook layout and the text will reflow to fit the new specifications.

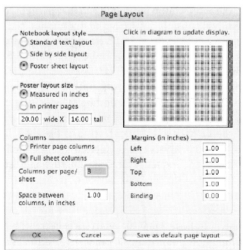

Figure 6-7.

As Notebooks grow in size, it becomes increasingly difficult to quickly find a particular location. Because of this, Gene Inspector allows the creation o. Bookmarks. Bookmarks can be attached to any object in a Notebook. A Bookmark menu lists all the Bookmarks in the Notebook and, when selected, will take the use immediately to that location in the Notebook. One use might be to place a Bookmark at the beginning of each day's worth of notes (or each Monday, or each month, or any other time point). To go to a certain date, then, it is a simple matter to select that date from the Bookmark menu. One might also indicate different projects in the Notebook with different Bookmarks to make jumping from one project to another a simple matter of selecting a menu item.

Rapid navigation can be extended to work between different Notebooks through the use of "aliases" (shortcuts). To use an alias, an object is selected and an alias o that object is created using a menu option. The alias can then be moved to a different location in the current Notebook, or it can be dragged (or copied and pasted) to a different Notebook. When the alias is double-clicked, it will take the user to the location of the original object used to create the alias, even if that object is in a different Notebook that is closed. Thus, it is possible to have a Notebook which contains protocols and then refer to that Notebook from any other Notebook by creating an alias to the original protocol.

Perhaps the most relevant type of object found in Gene Inspector Notebooks is the sequence analysis object. A sequence analysis object is created each time the user runs an analysis, as discussed in more detail in the next section. From the perspective of the Notebook, however, analysis output objects are similar to other Notebook objects in most ways — they can be selected, moved, copied, resized, edited, converted to appendices, and made to have background text flow around them. There are some properties that are unique to analysis output objects, however. The analysis output object has all the information needed to run that particular analysis such as the sequence(s) analyzed, styles used to display the output, and the parameters used for the analysis. Figure 6-8 shows a "Get Info" dialog for an analysis output object. I contains the basic information about the object, including how much disk and memory space the analysis object utilizes.

Any analysis object can also be used to rerun the exact same analysis. Double clicking the analysis object opens it up for editing. The editing can involve cosmetic changes such as changing the color or thickness of lines or the font, style and color of the axis labels. It can be used to change the title of the analysis as well. If desired the parameters used for the analysis can be altered and the analysis rerun – without the user having to reselect the sequences and starting the analysis over from scratch.

The last special feature of the analysis output object is that it can be "hotlinked" to the original sequence(s). Hotlinked analysis objects will contain a small indicator that tells whether the current analysis is up-to-date or not. A green dot in the upper right corner of the object indicates that the analysis is current, while a red and yellow symbol with an exclamation point indicates that the original sequence has changed since the analysis was last run. It is then possible to ask Gene Inspector to update the object and cause the analysis to be rerun using the parameters stored with the analysis output object on the updated sequence file. One interesting use of this feature is to create a Notebook with all the analyses you like to perform on a sequence and to have all of the output objects hotlinked to the sequence. The sequence file itself is really just a placeholder in this case, so that any time you would like to perform every analysis on a new sequence (using your predetermined parameters), all that has to be done is to replace the older sequence with the new one you want analyzed and then open up the Gene Inspector Notebook with all of the hotlinked analyses. The program will realize that the sequence has changed since the last run and will ask if you would like to update the analyses. Answering yes will update the entire Notebook of analyses using the new sequence.

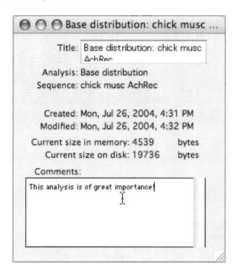

Figure 6-8.

6.5 Analysis Setup Panels

Analysis setups are the means by which analyses are created and run. The setup panel has a number of panes as shown in Figure 6-9. The three different panes can be viewed by selecting one of the three icons along the left. This figure shows the Input Sequences pane in which a single sequence, Dros hsp70, has been chosen. The "Range" section of the window allows either the selection of the whole sequence (the default condition) or a defined segment of the sequence. Pressing the "Add" button will add a sequence to this Analysis Setup by allowing the user to select the new sequence with the dialog shown in Figure 6-10. Note that since Gene Inspector can store multiple sequences in the same file, it is necessary first to select a sequence file (Drosophila HSPs as shown in the top part of the figure) and then to select one or more sequences from those available in the selected file (Dros hsp70 in this case). It is possible to select one or more individual sequences from one or more files using this dialog.

119

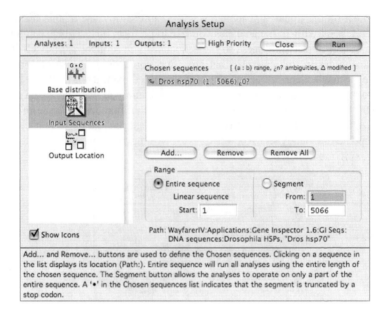

Figure 6-9.

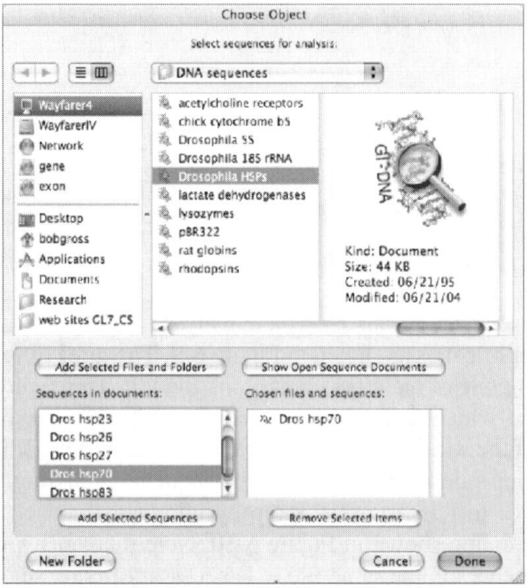

Figure 6-10.

Figure 6-11 shows the Analysis Setup Panel analysis parameters pane. This pane will look slightly different for different analyses because each analysis requires different parameters. The Base Distribution analysis setup pane is shown in this figure. It requires two numerical parameters (window size and offset, which are described in a diagram in the parameters pane) and a choice of bases using check boxes. This pane is specifying that the sequence should be analyzed using a sliding window of 50 nucleotides to calculate the total number of pyrimidines (C+T) in that window of 50. The window should be moved over 5 nucleotides and C+T calculated again, and the process repeated until the end of the sequence is reached.

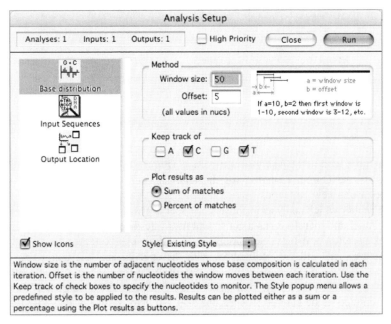

Figure 6-11.

The results of this analysis are shown in Figure 6-12. There are two pyrimidine rich stretches of sequence in this gene, one at about 1400 and the other at about 4600 nucleotides from the start of the sequence.

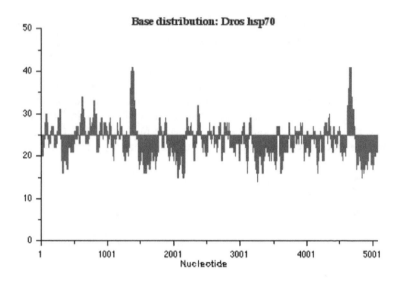

Figure 6-12.

The third icon in the Analysis Setup Panel (Figures 6-10 and 6-11) allows the user to specify the Output Location for the analysis output. This can be in any open Notebook or a new Notebook.

No matter which analysis is chosen, the user is always presented with an Analysis Setup Panel and therefore has a consistent user interface with which to interact. The Output Location and Input Sequence panes will always be the same. The Analysis Parameters pane will vary somewhat with the specific analysis being chosen, but there will always be an analysis parameters pane for the user to interact with when launching an analysis. At the bottom of every analysis pane is some text describing the parameters to be entered (see Figure 6-11) so help is always available if needed while not slowing down an experienced user.

The Analysis Setup Panel can contain multiple analyses. Using the Analysis Menu, it is possible to add additional analyses to an open Setup Panel. Analyses selected will be listed along the left side of the window, will have its own parameters pane, and will be performed on each of the sequences chosen in the input sequence pane. Thus, it is possible to perform multiple different analyses on multiple sequences using a single Analysis Setup Panel. If this set of analyses is something that the user feels will be repeated often and by others in the lab, the Analysis Setup Panel can be saved for future use. The user can name the Setup and it will be available from the Analysis menu. By carefully setting all the parameters for each of the individual analyses, and then saving the Setup, a knowledgeable user will make it possible for anyone else to reproduce the results. By having pre-defined analyses with parameters, others in the lab group can be confident that analyses of other sequences (using the saved Setups) will be comparable to the original analyses. One will not be in the position of trying to compare two different analysis runs on different sequences using different parameters, a situation that is far too common and is nearly impossible to interpret accurately.

6.6 Sample Analyses

Figure 6-13 shows the results of three hydropathy analyses on three different rhodopsins (bacteriorhodopsin, lamprey rhodopsin and *Xenopus rhodopsin*). A sliding window of 20 was used to calculate the average hydrophilic index using the table developed by Hopp and Woods[6]. This analysis portrays the degree of hydrophobicity along the length of a protein. Those regions that are most likely to reside on the surface of a protein are usually the most hydrophilic and the most antigenic. In this case, the three different species show some strong similarities in the profiles. The lamprey and the *Xenopus* are very close almost for their entire length but the bacteriorhodopsin, which is 100 amino acids shorter, shows some divergence in the middle of the sequence. In order to evaluate this more carefully, Gene Inspector can apply a process called median sieving (7). This will help to distinguish peeks of a given length; in this case about 20 amino acids that are needed to span a cell membrane.

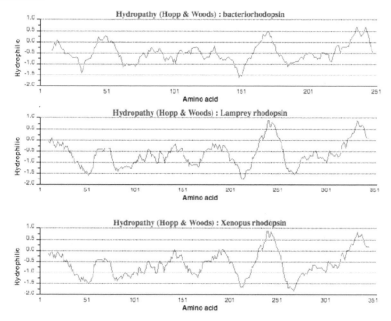

Figure 6-13.

After applying data sieving to these profiles, the results look like those in Figure 6-14. This "filtering" of the data leads to a much clearer comparison among the various rhodopsins and shows the strong similarities in the middle of the proteins as well as their ends. Median sieving can be applied to any sliding window output and can often help distinguish features that are not apparent from the raw data alone.

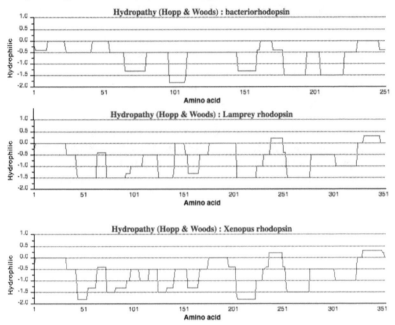

Figure 6-14.

One of the most important analyses available for comparing two sequences is the dot matrix analysis. This analysis gives a visual comparison between two sequences that the eye can pick up very easily, yet is hard to represent as a table or an actual sequence alignment. Figure 6-15 shows a dot matrix comparison of the DNA sequences for Drosophila hsp23 vs. hsp27. This plot places a dot at every location in which there is a match between the two sequences. The diagonal lines show regions where the two sequences share a stretch of similarity. Gaps in the diagonal lines show regions that have diverged.

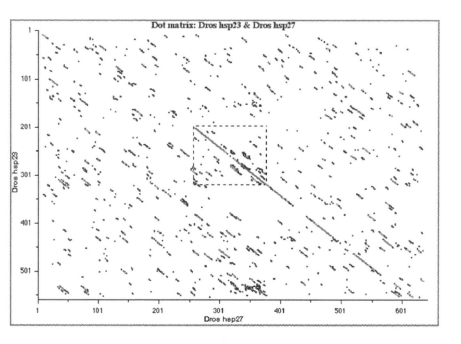

Figure 6-15.

In order to qualify for a match in this particular analysis, the different colored (shades) dots represent 9, 10, 11, or 12 matches out of a window of 15 nucleotides. Examining this plot shows a region in the middle of both sequences that looks to be interesting and very similar between the two sequences. The next obvious step, therefore, would be to do a sequence alignment of this region and assess the significance of the similarity shown in the dot matrix plot. This can be accomplished directly from within the dot matrix analysis output object by selecting with the mouse the region to be aligned. The dotted rectangle shows the selected region. After selecting the regions to be aligned, a menu choice starts the actual alignment, the results of which are shown in Figure 6-16. Notice the process here, in that a new analysis (a sequence alignment) was started from within an analysis output object in the Notebook. It was not necessary to write down the range of nucleotides to be used for each sequence and start setting up the analysis from scratch. Simply by selecting the region of interest in the dot matrix plot defined all the parameters needed to launch the new analysis.

Align 2 sequences (global): Dros hsp23 & Dros hsp27

First sequence: Dros hsp23
Second sequence: Dros hsp27
Scoring table: Nucleotide identity
Gap insertion penalty: 2.50
Gap extension penalty: 0.30
Unaligned ends treated as gaps.
Traceback: Upper path
Score: 94.90
Mean: 62.47
Standard deviation: 10.39
Z-score: 3.12
0 of 100 trial alignments scored greater than this one.

```
199   GG AAA GGATGGCTT CCAGGT CT GCAT G GAT GT GT CGCA CT T C G   241
      | |  | |  | | | | | | | | | | | | | |  | | | | |  | | | | | | | | | | |   | | |
256   GG C AAA GATGGCTT CCAGGT GT GCAT C GAT GT GT CGCA GT T C A   298

242   AGCCCA G CGA A CT GG T GGT C GGA GT GC AGGACAAC T CCGT GGT   284
      | | | | | |  | | |  | | |  . . .  | | |  . . .  | | |  . | | | | | | |  | | | | | | |
299   AGCCCA A CGA GCT GA C C GT C AAG GT GG T GGACAAC A CCGT GGT   341

285   GGT G GAGGG C AAC CAT GAGGAGC GC GAGGA T G ACC AT G   322
      | | |  . | | | | |  . | |  . | |  | | | | | | | | | | | | |  . | . | |
342   GGT A GAGGG GAAG CAC GAGGAGC GC GAGGA C G GCC - - -   376
```

Figure 6-16.

Figure 6-16 shows two aligned sequences and provides information about the significance of the alignment. As shown in the text at the top of the output object, 100 other alignments were run between these two sequence after scrambling the sequence of one of them. None of the 100 other alignments had a score as high as the current alignment of the actual sequences, indicating that important in the alignment (and that it was not due to an unusual base composition or some other artifact). The mean score of the scrambled alignments and the standard deviation are also given. This can be used to evaluate the significance of the alignment. The z-score is the number of standard deviations away from the mean that the actual alignment score represents. Higher z-scores indicate a more significant alignment.

Let's look at another example of the natural workflow that is encouraged by allowing new analyses to be launched from directly within an analysis output object. One of the earliest methods of looking for coding regions in a DNA is called Testcode[8] and it is still a very useful tool in many cases. The results of a Testcode analysis on Dros. hsp70 are shown in Figure 6-17. The top part of the figure indicates the likelihood that a region of DNA might be a coding region. There is a clear domain in the middle of the sequence that is likely to be coding. The bottom part of the figure contains two kinds of information. The arrows indicate open reading frames and the tick marks indicate the presence of rare codons (codons not used frequently in the organism chosen). Most coding regions for proteins that are abundantly produced do not contain many rare codons. Based on the high Testcode scores in the middle that corresponds to an open reading frame (in reading frame #3), and the paucity of rare codons in that open reading frame, it is highly likely that this open reading frame represents a true coding region.

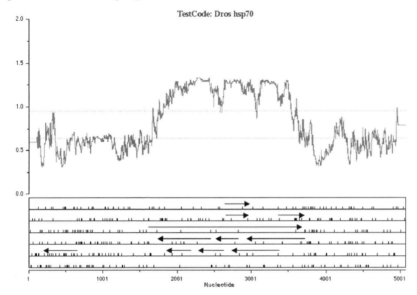

Figure 6-17.

Based on the results shown in Figure 6-17, the next thing that comes to mind is to examine the sequence that is coded for in this open reading frame and to perhaps do a BLAST search, which can be launched directly from within Gene Inspector. This is easily done without leaving the analysis output object. Double-clicking makes the object editable and the open reading frames selectable. Clicking on the appropriate open reading frame allows one to choose a menu item to translate the given range of DNA sequence. A new amino acid sequence is created and placed into a new sequence window. Thus, generating the relevant protein sequence directly from the Testcode output is a simple process.

6.7 Conclusions

Desktop sequence analysis tools have come a long way since their inception in the late 1970s and early 1980s. Analyses that were difficult to access and required experts to perform on mainframe computers 20 years ago are now done routinely on desktop computers. Further, programs like Gene Inspector provide a consistent and user-friendly interface that is easy to use for any biologist. It allows the researcher to follow his or her train of thought and follow up one analysis with the next logical analysis. The Gene Inspector Notebook allows all the analytical tasks to be annotated and discussed and can be used to store regular laboratory experimental data and discussions as well. The ability to save Analysis Setups with all parameters set appropriately means that anyone in a lab group can now perform analyses and compare results to others using the same saved Analysis Setups.

As genomic data becomes more readily available and gene expression data obtained through microarray experiments accumulates it is clear that we will need very sophisticated computational tools to begin to analyze the enormous amount of data being generated. Desktop computers are becoming more powerful every year and new and efficient algorithms are continually being developed. Although not practical now, it is very likely that in 5-10 years we will all be doing genomic scale analyses on our desktop computers, and probably places we never dreamed of on our laptops.

References

1. http://www.Textco.com/
2. http://informaxinc.com/content.cfm?pageid=13
3. http://www.dnastar.com/web/r3.php
4. http://www.accelrys.com/products/macvector/index.html
5. http://www.miraibio.com/products/cat_bioinformatics/view_dnasismax/index.html
6. Hopp, T.P. and Woods, K.R.: Prediction of protein antigenic determinants from amino acid sequences, PNAS (USA) 1981, 78(6):3824-3828.
7. Bangham, J.A.: Data-Sieving Hydrophobicity Plots, Anal. *Biochem.*1988,174:142-145.
8. Fickett, J.W.: Recognition of Protein Coding Regions in DNA Sequences, *Nucl. Acids Res.* 1982, 10(17):5303-5318.

Chapter 7

Repeat Mining: Basic Tools for Detection and Analysis

Adam Pavlicek, Oleksiy Kohany and Jerzy Jurka

Contact: Jerzy Jurka
Genetic Information Research Institute, 1925 Landings Drive, Mountain View, CA
E-mail: jurka@girinst.org

Abbreviations: CDS - coding sequence; ERV - endogenous retrovirus; HERV - human endogenous retrovirus; LINE - long interspersed element; LTR - long terminal repeat; MaLR - mammalian apparent retrotransposon; NA - nucleic acid; ORF - open reading frame; RR - Repbase Reports; RT - reverse transcriptase; RU - Repbase Update; SINE - short interspersed element; SSR - simple sequence repeat; TE - transposable element; TIR - terminal inverted repeat; TSD - target site duplication.

7.1 Introduction

Repetitive elements are DNA sequences present in multiple copies in a given genome. They are either generated *de novo* in the form of segmental duplications or tandemly arrayed repeats, or copied from some active master copies and integrated as interspersed repeats. Active copies are often considered to be genome parasites without a clear benefit to the host. Eukaryotic genomes contain several types of repetitive DNA created by various mechanisms.

Tandemly arrayed simple sequence repeats (SSR) consist of direct repetitions of the same consensus motif. Satellites, centromeric, and telomeric repeats represent a separate category of tandem repeats, being concentrated in well-defined chromosomal regions. Low copy repeats such as segmental duplications represent intra- and interchromosomal duplications. Interspersed repeats are mostly inactive copies of transposable elements (TEs). They contribute from a few percent to as much as 45% of eukaryotic genomes[1]. A separate category of repetitive elements are pseudogenes, nonfunctional copies of genes scattered across eukaryotic genomes. Pseudogenes originate either by genomic duplications (duplicated pseudogenes) or by reverse transcription of cellular RNAs (processed pseudogenes).

Due to their abundance and diversity, interspersed repeats are routinely analyzed during genome sequencing. Progress in genome sequencing, assembly and annotation including, but not limited to prediction of protein coding sequences, annotation of mRNA and regulatory elements, requires knowledge of genomic repetitive elements. Disregarding repetitive DNA can lead to serious errors in genome assembly and annotation. In this chapter we present a basic introduction to the structure of repetitive DNA and describe the main strategies and programs for detection of repeats, with emphasis on the main applications in genome assembly, annotation, and database searches.

7.2 Simple sequence repeats (SSRs)

Tandemly arrayed sequences or simple sequence repeats (SSRs), also known as VNTRs (variable number of tandem repeats), are rapidly evolving sequences that are ubiquitous in both prokaryotes and eukaryotes. They are formed by multiple repetitions of basic sequence motifs. SSRs are characterized by the *unit* sequence, unit *length* and the *number of units*. All parameters may vary significantly. Virtually all nucleotide combinations (patterns) are possible, but under- or overrepresentation of some variants is common within a genome, as well as between genomes of different species (see below). There is a wide range of unit sizes. SSRs are composed of units as short as 1 bp, but they may be as long as several hundred bp. Based on the unit length, SSRs are somewhat arbitrarily divided into micro- and minisatellites. *Microsatellites* are tandem arrays of short units, while *minisatellites* are composed of longer patterns. The boundary separating micro- and minisatellites considerably differs in the literature. Some sources define microsatellites as repetition of 1-6 bp long units, while others use higher limits (10-13 bp per unit) as a boundary between

133

micro- and minisatellites. The majority of satellites are relatively short (several tens to a few hundred base pairs), but depending on the species and chromosomal localization they can reach up to several dozens (or even hundreds) of kilobases. Traditional division of tandem repeats into microsatellites and minisatellites is based on early hypotheses about the mechanism of their expansion. Replication slippage during DNA amplification was believed to be the underlying phenomenon of microsatellite expansions[2,3], while minisatellites were thought to be products of unequal recombinations[4,5]. However, currently there are several potential meiotic and/or mitotic mechanisms for the expansion or contraction of both micro- and minisatellites. They range from slippage during replication and gap repair, to recombination and gene conversion[see 6-10 and refs. therein].

SSRs are abundant in noncoding regions across several eukaryotic clades, except for tri- and hexanucleotide repeats prevailing in protein coding regions[11-13]. In contrast to primates, several studied plant genomes contain fewer tandem repeats derived from TEs and show increased density of satellites in untranslated regions[13]. In comparison with other taxonomic groups, primates contain an excess of homonucleotide repeats, particularly A and T tracts, probably due to contribution of polyA tails by non-LTR-retrotransposons such as LINE1, Alu, and processed pseudogenes[12]. In summary, the distribution and abundance of different tandem repeat types significantly varies within and between genomes[8,10-12,14].

7.3 Interspersed Repeats

There are two major classes of transposons in the eukaryotic genomes: Class I includes retrotransposons, replicating through RNA intermediates, and class II - DNA transposons. Both classes contain replication-competent *autonomous* transposons, and *nonautonomous* elements, parasiting on the replication machinery of autonomous repeats. The proportion of nonautonomous to autonomous elements depends on the class of repeats and the genome under consideration. In addition to proteins encoded by *autonomous* elements, both autonomous and nonautonomous interspersed repeats depend on various factors produced by the host cell for their proliferation.

Retrotransposons amplify via an RNA intermediate, reversely transcribed into DNA and integrated into the genome. They depend on reverse transcriptase (RT) enzyme encoded by active elements. Eukaryotic genomes harbor two major groups of retro(trans)posons: *non-LTR retrotransposons* (also known as *polyA-retrotrans-posons*) - LINEs, SINEs, processed pseudogenes, and *LTR-retrotransposons* similar to retroviruses.

Non-LTR-retrotransposons represent a widespread class of mobile elements found in various eukaryotes [15-17]. They lack long terminal repeats (LTRs) and contain polyA, or other tails composed of specific simple repeats. Unlike in retroviruses, RNA intermediates of non-LTR-retrotransposons are structurally collinear with integrated DNA and explore internal promoters (Figure 7-1). There are two major classes of non-LTR-retrotransposons: autonomous *long interspersed elements* (LINEs) and nonautonomous *short interspersed elements* (SINEs), and retroposed copies of cellular RNAs - processed pseudogenes (retropseudogenes).

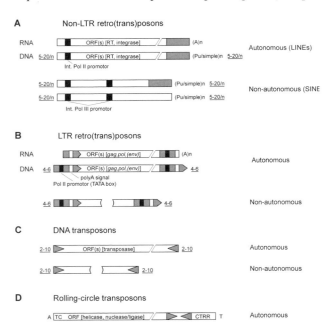

Figure 7-1. Genome structures of main eukaryotic transposons.
Schematic structures of the major transposons classes. For autonomous retrotransposons (A,B) the organization of both RNA and DNA sequences is shown and for other elements only integrated DNA is drawn. Internal deletions frequently found in nonautonomous elements are depicted. Specific positions of promotors in retrotransposons are marked by black boxes. Position of the polyA signal in terminal repeats of LTR-retrotransposons is highlighted by white boxes. Gray boxes at the end of LINE and one SINE element represent 3' homologous regions, frequently shared by LINEs and SINEs (A). Gray boxes in LTR-retrotransposons (B) correspond to sequences of long terminal repeats (LTRs) in DNA. Small gray triangles show position of terminal inverted repeats (TIRs) in DNA transposons (C), and positions of the hairpin structure composed of two short inverted repeats at the 3' end of rolling circle transposons (D). Underlined values flanking interspersed repeats show the sizes of target site duplications (TSDs) followed by "/n" to indicate lack of TSDs in some LINEs. PolyA and other simple repeats are marked as well. Rolling circle transposons are flanked by 5' A and 3' T nucleotides, corresponding to the characteristic target of these transposons (D).

LINEs can be classified into two major subtypes with several subgroups[16]. One subtype is characterized by the existence of a restriction-like endonuclease domain and a single open reading frame (ORF). The other type of LINEs, including human LINE1 (L1) elements, harbor an apurinic/apyrimidinic-like endonuclease encoded by one of its two ORFs. The human genome contains three major families of LINEs. Two of them, LINE2 and LINE3, homologous to CR1 (chicken repeat 1), are inactive. The third one, LINE1, still remains active in mammals.

LINE1 transcription is initiated from a poorly characterized internal pol II promoter[18,19]. LINE1 retrotransposons carry two open reading frames (ORFs) encoding ORF1p and ORF2p proteins. The second ORF (ORF2) encodes a multifunctional protein with an endonuclease, reverse transcriptase, and cysteine-rich terminal domain[20,21]. The biological function of the ORF1p proteins is less understood, except in that they nonspecifically bind single-stranded mRNA/DNA molecules[22] and have nucleic acids chaperone activity[23]. Surprisingly, a conserved esterase domain, and plant homeodomain, were recently reported in ORF1 of CR1-like elements[24], the most widely distributed clade/superfamily of non-LTR retrotransposons. This suggest that CR1-like ORF1p proteins are either lipolytic enzymes or are involved in protein-protein interactions related to chromatin remodeling[24].

LINE1 retrotransposons show a target preference for TTTTAA insertion motifs[25]. During the insertion, target site duplications (TSD) 5-20 bp long are created with a maximum of 15-16 bp[25]. Occasionally, short deletions within target sequences may occur[26]. Specific target motifs were reported for some other LINE classes as well [27,28]. However, LINE2 and CR1-related LINEs are typically not flanked by TSDs.

SINEs are 100-400 bp long nonautonomous elements without any protein coding capacity. The major class of SINEs includes elements derived from tRNA molecules. The second minor SINE class originated from 7SL genes and includes human Alu and rodent B1 elements. The third class are 5S rRNA-derived elements from the zebrafish genome called SINE3[29]. All SINEs exploit internal RNA pol III promoters[29,30].

The amplification of SINEs is thought to be dependent on LINEs. Recently, experimental in vitro assay demonstrated that an eel SINE can be mobilized by the retrotransposition machinery of LINEs[31]. SINEs are very well adapted parasites of LINEs and many even share 3' sequences with corresponding LINE counterparts[32]. Each SINE is specialized for a particular type of LINE and extinction of a LINE family is without exception followed by extinction of its SINE parasites [33,34]. Human Alu and processed pseudogenes share the same insertional characteristics as LINE1, including 5-20 bp long TSDs and the TTTTAA target motif [25].

The human genome contains three groups of SINEs: extinct MIR, MIR3 that coamplified with LINE2 and LINE3 (CR-1), and still active Alu sequences thought to be mobilized by LINE1. MIR and MIR3 are tRNA-derived elements[35,36], while Alus are dimers of 7SL RNA, a part of the RNA scaffold of the ribosomal signal recognition particle[37].

A second group of retrotransposons includes long terminal repeats (LTR)-retrotransposons, resembling integrated proviral genomes of retroviruses. They form several major subgroups including well-studied vertebrate endogenous retroviruses (ERVs). Around 8% of human DNA is derived from LTR-elements[1]. They are likely to originate from ancient retroviral infections transmitted to germline cells. Human endogenous retroviruses (HERVs) show similarity to coding regions of exogenous retroviruses (Figure 7-1) and comprise parts of LTR-elements[see 36,38-40]. Non-autonomous LTR elements such as mammalian apparent LTR retrotransposons (MaLRs)[41], are without protein coding capacity and depend on autonomous LTR-elements, analogously to SINE dependence on LINEs. LTR retrotransposons are flanked by LTRs that contain all necessary transcriptional and regulatory elements.

Proliferation of LTR-retrotransposons follows the same mechanism as used by retroviruses with reverse transcription occurring in a cytoplasmic virus-like particle, primed by a tRNA. This is in contrast with the nuclear location and targeted DNA priming of LINEs. LTR-retrotransposons are flanked by short (4-6bp) TSDs, and typically contain TG and CA dinucleotides at provirus termini. Autonomous LTR elements contain gag, pol, (pro) genes (Figure 7-1), which encode structural capsid proteins, protease, reverse transcriptase, RNAse H, and integrase. The number of open reading frames (ORFs) ranges from a single ORF for many Bel and Copia-like elements, to several ORFs, particularly in Gypsy related retrotransposons. Some families also contain ORFs homologous to envelope proteins and/or additional accessory genes[42; see also ref. 36 and references therein]. Nonautonomous LTR elements have obligatory LTR-like structures and primer binding sites, required for retroviral reverse transcription. Moreover, they may or may not share similarity to internal, protein-coding regions (Figure 7-1).

Eukaryotic DNA transposons resemble bacterial transposons. They contain terminal inverted repeats (TIRs) and encode a transposase that mediates mobility through excising a transposon at both its termini and subsequently inserting it into a new position (cut-and-paste mechanism)[1,33]. Replication of DNA transposons is accomplished by the cell replication machinery during replication of the host DNA. DNA transposons found so far in eukaryotic genomes are characterized by structural hallmarks including terminal inverted repeats (TIRs) and 2- to 10-bp target site duplications (TSDs), generated by staggered endonucleolytic cleavage of the target DNA by their transposases[36]. Similarities between different transposons are often limited to a 3 amino-acid motif, insufficient to show homology. Nonautonomous DNA transposons do not encode functional transposases. However, they share partial similarity to automonous copies. The simplest of them include only TIRs essential for recognition by heterologous transposase and amplification.

Rolling circle (RC) transposons represent a special class of DNA transposons. The first eukaryotic RC elements termed Helitrons were detected in genomes of *A. thaliana*, and subsequently in *C. elegans* and *O. sativa*[43]. They encode a 5'-to-3' DNA helicase and nuclease/ligase similar to those encoded by known bacterial rolling-circle replicons. Helitron-like transposons contain conservative 5'-TC and CTRR-3' termini and unlike cut-and-paste DNA transposons they lack TIRs. The

137

rolling circle transposons contain 16- to 20-bp hairpins separated by 10-12 nucleotides from the 3'-end and transpose precisely inside the 5'-AT-3' dinucleotides without TSDs. Analogously to other groups of transposons, rolling circle transposons include nonautonomous elements. They share common 5'-TC and CTRR-3' termini, TA-target specificity and other structural hallmarks with autonomous Helitrons, but they do not code for any complete set of proteins encoded by the autonomous elements[43].

7.4 Reference collection of repeats

The first reference collection of 53 separate human repetitive elements was published in 1992 and named 'Repbase'[44]. A fast growing knowledge of TEs from other eukaryotic species led to expansion of the originally human-limited database to a database of other eukaryotic repetitive elements called Repbase Update (RU)[45]. Currently, the RU database is released in 3 different formats: the EMBL format contains detailed description and annotation of every RU consensus sequence. The Fasta format provides RU in the common format used in biological sequence analysis. In addition, RU has a special section of Fasta libraries tailored for the RepeatMasker detection software. RU is updated monthly but its RepeatMasker version is updated less frequently.

RU became *de facto* the main source of eukaryotic repetitive sequences and is widely used in genome assembly and genome annotations including the human and mouse genome projects[1,46]. To facilitate proper referencing and documentation of individual contributions, a peer-reviewed electronic journal entitled Repbase Reports (RR) was established in fall of 2001. Similarly to RU, RR is published monthly and available from the Genetic Information Research Institute website at http://www.girinst.org.

Repbase Update contains several sections, including genome-specific collections of repeats and several general sections for TEs shared by rodents, vertebrates, invertebrates and plants. Other sections include processed mammalian pseudogenes (pseudo.ref) and simple repeats (simple.ref). Every sequence entry in the EMBL format contains several fields (Figure 7-2), describing the source (species), date, and authors of the submission, references to published literature (if any), basic family description, and the DNA sequence. Recent submissions usually include protein translations of ORFs for autonomous elements. The translated protein sequences are very useful for identification of transposable elements in newly sequenced genomes using protein-based similarity searches (see section 7.5.2.2).

```
ID    HELITRON1_AG repbase; DNA; INV; 6666 BP.
CC    HELITRON1_AG DNA
XX
AC    ;
XX
DT    11-NOV-2002 (Rel. 4.1, Created)
DT    11-NOV-2002 (Rel. 4.1, Last updated, Version 1)
XX
DE    HELITRON1_AG, a rolling-circle DNA transposon - a consensus
DE    sequence.
XX
KW    DNA transposon; HELITRON class; Rep/helicase;
KW    HELITRON1_AG.
XX
OS    Anopheles gambiae str. PEST (consensus)
CC    consensus
OC    Eukaryota; Metazoa; Arthropoda; Hexapoda; Insecta; Pterygota;
OC    Neoptera; Endopterygota; Diptera; Nematocera; Culicoidea;
OC    Anopheles.
XX
RN    [1]
RP    1-6666
RC    [1]  (bases 1 to 6666)
RA    Kapitonov V.V., Jurka J.;
RT    "HELITRON1_AG, a rolling-circle DNA transposon from
RT    African malaria mosquito.";
RL    Repbase Reports 2:(10) p. 8 (2002)
XX
CC    HELITRON1_AG is the first example of HELITRONs (a rolling-circle
CC    DNA transposons) present in insects. The A. gambiae genome
CC    contains many families of HELITRONs. HELITRON1_AF is one of them.
CC    A defective copy of a HELITRON-like element is also present
CC    in the D. melanogaster genome (AE002840, positions 9827-9264).
CC    HELITRON1_AG has all basic hallmarks of HELITRONs: it is
CC    flanked by 5'-TC and CTAG-3', is inserted into the AT target
CC    site, and encodes a 1792-aa HELITRON-like Rep/Helicase protein,
CC    called AGHEL1p.
CC
CC    AGHEL1p (positions 820-6195):
CC    MQPSVSDIVSQNLTPPEETPNEKKARLQRKRQALYRAKKRLPGAAPTLAAVQDDQQAVPSTLAGSSLSAA
...
XX
CC    [1] (Consensus)
XX
SQ    Sequence 6666 BP; 1706 A; 1761 C; 1639 G; 1560 T; 0 other;
      tctatatata aattttcgtt aacgctgacg ctcgatgtca cactttgcgt tacgctgttt        60
...
```

Figure 7-2. Representative example of a recent RU entry in the EMBL format.
The entry contains unique identification (ID) HELITRON1_AG, composed of 3 parts - class of elements (HELITRON), followed by identification number "1" in species "A" - Anopheles gambieae). Other fields include date of submission/update (DT), keywords (KW), taxonomical information about the host species (OC), reference fields (RN-RL), description of biological properties, replication mechanism including translation of ORFs (CC), and base composition of the consensus sequences (SQ), followed by the sequence itself.

7.5 Detection of repetitive elements

7.5.1 Detection of SSRs

The primary objects of SSR studies are short microsatellites (2-6 bp), due to their abundance, relatively uniform chromosomal distribution, relationship to human neurodegenerative diseases, and usefulness for genomic mapping and population studies. The spectrum of available programs for SSR detection reflects the bias for detection and analysis of satellites with short units (Table 7-1).

Program (package)	Availability	System	Reference	URL / Description
Tandem Repeat Finder	B, W	Linux/Unix Mac OS X Win/DOS	50	http://tandem.bu.edu/trf/trf.submit.options.html
				An excellent program to locate and display tandem repeats in DNA sequences. Repeats with pattern size in the range from 1 to 2000 bases are detected.
satellites	W	- (web-only)	79	http://bioweb.pasteur.fr/seqanal/interfaces/satellites.html
				Web-based interface for detection of satellites and periodic repetitions in biological sequences.
Sputnik	B, S	Linux/Unix Mac OS X Win/DOS	Abajian, C unpublished data	http://espressosoftware.com/pages/sputnik.jsp
				Scans for 2-5 long tandem repeats.
EMBOSS	S	Linux/Unix Mac OS X	51	http://www.hgmp.mrc.ac.uk/Software/EMBOSS/index.html
				Etandem looks for tandem repeats in a nucleotide sequence. Running with a wide range of repeat sizes is inefficient, but it's possible to use equicktandem to give a rapid estimate of the major repeat sizes. There are two other related programs in EMBOSS: einverted finds DNA inverted repeats, palindrome looks for inverted repeats in a nucleotide sequence.
Censor on-line	W	- (web-only)	48	http://www.girinst.org/Censor_Server-Data_Entry_Forms.html
				The on-line version of Censor identifies SSR using de novo detection of SSR, including complex motifs composed from several tandem repeats.

Table 7-1. Software for detection of simple sequence repeats. Availability: B - binaries, S - source, W - web interface. FreeBSD-based Mac OS X is listed separately, but as a matter of fact it is also a UNIX-related system.

There are two basic strategies for identification of tandem repeats: *de novo* identification and similarity-based detection. *De novo* identification is independent of any previous knowledge of SSR and virtually all possible combinations of nucleotides in satellites can be detected in a new sequence. Table 7-1 lists some of the most popular programs for *de novo* SSR detection.

The similarity-based detection of tandem repeats (and repetitive sequences in general) requires comparison with known sequences of repetitive elements, i.e. requires databases of SSRs. Censor, RepeatMasker and MaskerAid programs, designed for detection of both interspersed and tandem repeats, can be used for identification of mainly short SSR. They use SSR library simple.ref from Repbase Update[47]. The web version of Censor also runs *de novo* analysis of SSRs based on the previously reported algorithm[48]. The programs are described in detail in the section dedicated to interspersed repeats. These programs, however, are unable to detect unknown SSR types mainly including rare tandem repeats and/or satellites with long units. As mentioned in the above section, abundance of different SSR types greatly varies between different species. For new genomes or studies of long-unit minisatellites it is therefore better to use *de novo* detection programs.

Some genome-wide *de novo* analyses of short SSRs by different research groups use independent in-house programs[see for example ref. 12,14]. While *de novo* detection of short minisatellites is relatively easy (very often by unaided eyes), repeats with long or mixed patterns are much harder to detect[49,50]. Many *de novo* detection programs find only the most abundant microsatellites, but not minisatellites, especially those with larger unit sizes. Also, scanning for SSRs of various unit sizes is both processor and memory intensive in comparison with the detection of short microsatellites. For longer minisatellites it is possible to use Tandem Repeat Finder [50] or Etandem program from the EMBOSS package[51]. Etandem is slow if a wide range of SSR sizes is detected. On the other hand, Tandem Repeat Finder[50] seems to be an excellent solution, since it combines high sensitivity and the ability to detect SSRs of various sizes ranging from 1-2000 bp with reasonable time and memory requirements. A laptop computer with a mobile Pentium 4 processor, 512 MB of RAM, running Linux (kernel 2.4.18-3), scanned 33Mb of human sequence data for SSRs with units 1-2000 bp long in 14 minutes, using approximately 165 MB RAM (command line parameters: 2, 2, 7, 80, 10, 50, 2000). Of course, detection of short microsatellites in smaller chunks of DNA sequence is much faster.

In conclusion, for detection of major SSR types or repeat masking, especially in well-studied genomes it is sufficient to use standard repeat detection programs like Censor, MaskerAid or RepeatMasker based on Repbase Update libraries. For studies of long-unit SSRs, especially in poorly characterized genomes, it is better to use *de novo* detection programs such as Tandem Repeat Finder.

7.5.2 Detection of interspersed repeats

Given the importance of proper repeat identification for genome assembly, gene prediction, as well as structural and functional annotation of DNA and RNA sequences in general, it is not surprising that there are several specialized tools for repeat detection and masking (Table 7-2). Analogously to the analysis of tandem repeats, the analysis of interspersed repeats also employs two basic approaches: *de novo* detection, independent of the previous TE knowledge and similarity based methods.

Program (package)	Availability	System	Reference	URL / Description
Censor	S, W	Linux/Unix Mac OS X	56; Kohany et al.: unpublished data	http://www.girinst.org/Censor_Server-Data_Entry_Forms.html
				General package for identification of repetitive DNA in both nucleic and amino acid sequences. Requires Wublast and external repeat libraries.
RepeatMasker	S, W	Linux/Unix Mac OS X	Smit AF, GreenP: unpublished data	http://www.repeatmasker.org/
				General package for identification of repetitive DNA in nucleic acids sequences. Requires Cross_match and external repeat libraries.
MaskerAid	S	Linux/Unix Mac OS X	58	http://blast.wustl.edu/maskeraid/
				Program performs a Wublast based detection of repetitive sequences. Can be used with RepeatMaker for fast repeat identification. Requires Wublast and external repeat libraries.
Recon	S	Linux/Unix Mac OS X any OS with Perl and C	52	http://www.genetics.wustl.edu/eddy/recon/
				The RECON package performs de novo identification and classification of repeat sequence families from genomic sequences.

Table 7-2. *Software for detection of interspersed repeats. Availability: B - binaries, S - source, W - web interface.*

7.5.2.1 *De novo* detection

An intuitive way to detect unknown transposable elements is based on their basic property - presence in multiple copies. Cross-genome similarity searches reveal any multicopy DNA sequences including transposable elements (TEs). Typical programs used for *ad hoc* intergenome comparisons are the same as those used in general database searches such as BLAST, Fasta, Cross_match, etc. Simple manual inspection of particular DNA segments and their genomic homologues can detect the most abundant types of TEs. However, if the goal is to detect a large number of interspersed repeats, automatic procedures are required. The most direct way is to apply general clustering algorithms that produce clusters based on pairwise similarities between elements detected by similarity searches. One may use results from the genome similarity searches using BLAST or Fasta and perform a cluster analysis using either alignment scores, percentage identity, or percentage of positive nucleotides. The pairwise similarity/identity values can be processed through a variety of clustering methods used in phylogenetic analyses like UPGMA, neighbor-joining, etc. The resulting 'similarity' trees can guide separation of repetitive elements into distinct families.

The main drawback of simple clustering methods is the inadequate treatment of repeat fragments. If two different genomic elements are derived from the same family of transposons, but from different parts of the transposon sequence, no relationship between these two fragments will be detected. In other words, the standard approach of single linkage clustering of local pairwise alignments is primarily useful for abundant or very young repetitive families with many complete elements. Reconstruction of complete consensus sequences from fragmented elements requires more sophisticated methods.

An interesting attempt in this direction is Recon[52], searching the boundaries of individual copies of the repeats and trying to distinguish homologous but distinct repeat families. Recon doesn't work with actual sequences, but it processes pairwise similarities from other programs such as BLAST and Fasta, and uses overlaps between pairwise blocks to reconstruct the complete sequence. Although the program represents a significant advance in automated *de novo* identification of transposable elements, it cannot solve some biological intricacies as admitted by the authors themselves. The program is sensitive on highly rearranged elements such as endogenous retroviruses with internal portions removed by LTR-LTR recombination [52]. Moreover, obtained consensus sequences may be rearranged. *De novo* detections are also not very effective for low-copy and older repeats. Also such an approach may misidentify frequent genomic duplications, recent multigene families, or transduction of genomic DNA by LINEs as repetitive sequences. Finally, although the automatic clustering can produce consensus sequences of TE families, it does not give any biologically relevant information about them. Additional analysis is necessary to obtain information about the replication mechanism, coding capacity, etc.

Another approach in *de novo* detection can be based on structural characteristics of TEs[53]. The algorithm described by the authors and implemented as LTR_STRUC identifies and automatically analyzes LTR retrotransposons in genome databases by searching for structural features characteristic of such elements.

In conclusion, specialized clustering programs such as Recon are useful for a first-pass automatic classification of repeats in newly sequenced genomes. The resulting consensus sequences can be used for repeat masking in gene prediction and other annotations. To obtain more accurate consensus sequences and biological information about them, additional filtering and processing is needed. Finally, it should be stressed that *de novo* detection programs serve for identification of new repeats and their consensus sequences; they do not mask query sequences.

7.5.2.2 Similarity based detection

The principle of similarity-based approaches is to compare a DNA sequence with libraries of known repetitive sequences. There are several programs for similarity-based detection but all use the reference collections of repetitive sequences deposited in Repbase Update as a primary source of reference repeats. In general, any custom library can be used with these programs to detect any sequences of interest.

The repeat detection programs discussed below depend on external routines for a similarity search, namely Wublast[54] or Cross_match[55].

Censor

Censor is the first program for repeat detection released in 1994 and published in 1996[56]. Censor is composed of Perl and C++ modules for identification of both interspersed and tandem repeats using similarity searches. The original version of Censor is already outdated, but an upgraded version continues to be available as an on-line server at http://www.girinst.org/Censor_Server.html (Figure 7-3). It analyzes DNA/RNA sequences for repeats and provides a detailed description of repetitive elements from Repbase Update annotation. A new portable version of Censor is available at http://www.girinst.org/server/censor/ [Kohany et al., unpublished data]. Censor can be run on symmetric multiprocessor machines. It can be installed on virtually any system with Perl and a C++ compiler, but due to dependence on the strictly unix Wublast program[54], it can only run on Unix/Linux systems (including FreeBSD-based Mac OSX) with Wublast installed.

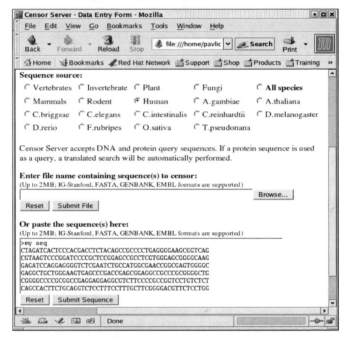

Figure 7-3. Censor web entry form.
The first field (box) offers various species options (note, both Censor and RepeatMasker are only able to detect repeats homologous to those present in RU or custom database). The entry sequence should be in the Fasta, IG/Stanford, GenBank, or EMBL format. It can be pasted into the form or loaded from a local file. After analysis, the results are displayed in the user's web browser. For all detected elements, the RU annotation is provided.

Censor uses the Wublast package to detect similarity between repeat libraries and nucleic acid (NA) sequences. For simple NA-NA searches, Blastn can be applied (default). For more sensitive detection of distantly related protein coding sequences, a 6-frame search (using tBlastx) or protein-NA search (tBlastn) is available. Censor automatically chops long sequences into smaller fragments and subsequently reduces Wublast memory requirements. The program reports positions of repetitive elements in 'map' files (Figure 7-5a); 'masked' files contain repeat-masked sequences, and individual repeats sequences are stored in 'found' files. General information on the query sequence(s) and repeat content is stored in 'tab' files. There is also a possibility to obtain alignments of repetitive sequences with the consensus using the -aln function.

RepeatMasker

RepeatMasker (Smit AF, Green, P: unpublished data) is a popular program for identification of tandem as well as interspersed repeats. The program is written in Perl and uses Cross_match[55] to run the Smith-Waterman search[57]. There are two web servers (Figure 7-4), but a local version of the program is also available upon request. RepeatMasker can run on multiprocessor machines and on computer clusters. Similarly to Censor using Wublast, RepeatMasker is restricted by Cross_match to Unix-related systems. The program uses slightly modified, customized repetitive libraries, available at the Repbase Update site. Due to these modifications, the RepeatMasker nomenclature is similar, but not 100% compatible with the basic RU libraries. Manual modification of repeat libraries prevents frequent updates and hence causes some delays in releases of customized libraries.

The program is limited to nucleic acid sequence comparisons only. Like Censor, RepeatMasker produces masked sequences (masked files), repeat maps (out files; Figure 7-5b), alignments if requested (-aln switch), and a summary table of repeat content of analyzed sequences (tbl files). In addition to simple identification of repetitive fragments, a defragmentation algorithm implemented in the program attempts to reconstruct more complete elements from genomic fragments (Figure 7-5b). Such repeat fragmentation is typical for old vertebrate elements with many indels.

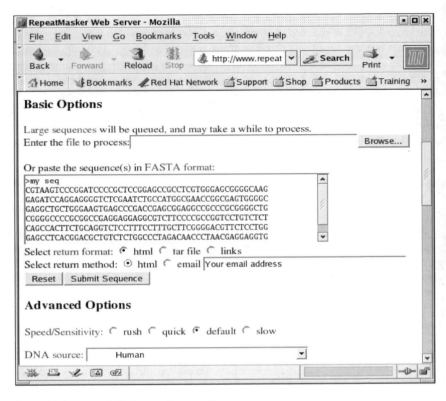

Figure 7-4. RepeatMasker web entry form.
The form starts with sequence entry: as in Censor, it can be from a local file or pasted using the clipboard. The only accepted format is Fasta format. Return format may be in html (for web browser) with repeat map, masked sequence in Fasta format, and summary table. Compressed (gzip) tar archives are suitable for local download, and contain the repeat map (out file) and masked sequences in Fasta format (masked file). Alternatively, it is possible to select the email return method instead of html. Again, there are available species options and several output options below.

A

```
1       2    3          4      5     6    7    8     9    10     11
seq     start end       family start end   ori  ident pos  score  class
query   167 184         (AG)n  1     18    c    0.91  0.91 120    simple
query   208 235         (GAAAA)n 2   29    c    0.91  0.91 177    simple
query   497 764         LTR1   27    288   d    0.79  0.79 1128   LTR
```

B

```
1    2    3    4    5      6            7    8      9   10       11          12       13   14     15
SW   perc perc perc query  position in query      matching repeat           position in  repeat
score div. del. ins. sequence begin    end  (left) repeat class/family begin end  (left) ID


3072 21.5 10.2 0.9  query  421005 421815 (78185) + L1ME1   LINE/L1       5078 5812 (532)  31
1659 22.9 0.3  4.2  query  421816 422134 (77866) + AluJo   SINE/Alu      1    307  (5)    32
2233 10.4 0.0  0.0  query  422138 422424 (77576) + AluSx   SINE/Alu      1    287  (25)   33
3072 21.5 10.2 0.9  query  422425 422493 (77507) + L1ME1   LINE/L1       5813 5998 (346)  31
```

Figure 7-5. Examples of sample output maps of repetitive elements from Censor and RepeatMasker.
(A) Censor stores the maps in 'map files'. The map files contain the following columns (from left to right): (1) name of the query sequence, (2) start in query, (3) end in query, (4) repeat family, (5) start at the consensus, (6) end at the consensus, (7) strand, (8) mean identity to the consensus, (9) mean proportion of positive (similar) positions, (10) Wublast score, (11) class of repeat.
(B) RepeatMasker, including the MaskerAid version saves maps into 'out' files. The 'out' files contain the following columns (from left to right): (1) Smith-Waterman score from Cross_match, (2) percent of divergence from the consensus, (3) percentage of deletions, (4) percentage of insertions, (5) name of the query sequence, (6) start at query, (7) end at query, (8) remaining bps at the 3' end of element, (9) orientation, (10) repeat family, (11) class / subclass of repeat, (12) starting position at the consensus, (13) end at the consensus, (14) remaining bps in the consensus, (15) identifier of repeat. The last column provides unique identifier for independent insertions using RepeatMasker's defragmentation algorithm. The Cross_match search identified 4 fragments (2 LINE1 and 2 Alu fragments). Looking at consensus positions, one can see that the first fragment ends at position 5812 at the L1ME1 consensus sequence, and the second LINE starts immediately after at position 5813. This indicates one old L1M LINE element with 2 Alu insertions at position 5812. In such situations RepeatMasker recognizes two fragments as one element and the identification number (ID) of fragments is the same - 31 in our case.

MaskerAid

Probably the main drawback of RepeatMasker is the extensive processor and time requirement for running a rigorous Smith-Waterman alignment[57] implemented in Cross_match. MaskerAid[58] was created as an accelerator of RepeatMasker, replacing Cross_match by fast Wublast. The resulting search is several times faster than RepeatMasker with only a slight decrease of sensitivity.

Like Censor and RepeatMasker, it requires the unix platform. It can be run as a stand-alone program or it can be invoked by the -w parameter in RepeatMasker version June 2000 or higher. The main goal of MaskerAid is fast and efficient identification of repetitive DNA and masking. Some advanced options such as generation alignments between detected elements and the library sequences (usually consensus sequence) are not supported.

7.5.2.2 Rearrangements of interspersed repeats and issues with detection of complete elements

The majority of transposons in eukaryotic genomes are defective, i.e. they contain point mutations, indels and structural rearrangements. Large sequence rearrangements or insertions disrupt homologies between genomic elements and their consensus sequences. As a consequence, similarity search programs (Wublast, Cross_match) detect separate fragments instead of a single element. To reconstruct TE, a defragmentation (concatenation) of fragments is required. Knowledge of the rearrangement mechanisms is necessary to separate random multiple insertions from several fragments derived from single insertions, and therefore for design of automatic defragmentation methods.

Apart from obvious mutations such as insertions of repeats into repeats and random deletions, there are many family-specific variations. Typical examples include 5' truncations and inversions in human LINEs[59], or LINE1 3' transductions [60-62]. LTR-retrotransposons are affected by recombinations via template switching during retroviral reverse transcription[63], splicing[64,65], and amplification of TEs by an enzymatic machinery of other repeat class[64-66]. SINEs, LINEs, and processed pseudogenes often contain AT-rich microsatellites[12,65,67,68]. Many 3' SINEs tails resemble 3' ends of LINEs[32], and they can be misidentified as LINEs or vice versa. CpG-rich repeats in methylated genomes such as human Alu elements are the subject to fast CpG decay[33,69,70]. 85-90% of human endogenous retroviruses are solo LTRs formed by internal homologous recombination between terminal LTRs[1,33,71,72].

All these and other rearrangements can complicate identification of complete TE elements. In the ideal situation, alignment between a genomic element and its family consensus sequence forms a single block of similarity. Sequence rearrangements listed above can lead to insertion of gaps in the alignment, or even to detection of separate fragments. Proper defragmentation should take into account at least the following parameters: (1) density of repeats from a particular class in given regions,

(2) distance between two fragments derived from the same family, after removing other TE insertions, (3) size of deletions or overlaps, if any, in the corresponding query consensus sequence and (4) orientation of fragments.

There are no simple rules for defragmentation. In the best case scenario, closely spaced fragments, after removing insertion of unrelated TEs and SSR expansions, should retain the same orientation. If they form a continuous alignment with minimal gaps, they are likely to be derived from the same element. However, this simple rule fails in many instances. While the requirement for the same orientation works for the majority of repeats, in the case of human LINEs it would report LINEs with 5' inversion as two adjacent, but independent insertions in the opposite orientation. Many 5' inversions are accompanied by internal deletions or duplications, so one should also allow for indels in the alignment. A more detailed description of all such problems and the corresponding solutions is out of the scope of this chapter and will be discussed elsewhere.

In summary, a detailed knowledge of major rearrangements is necessary to design an accurate method for the defragmentation of TEs. However, given the diversity of proliferation mechanisms and genomic organizations of repetitive elements, there are no simple solutions for every situation. In particular, endogenous retroviruses undergo an extremely complex spectrum of rearrangements and need special treatment. For example, the database of human endogenous retroviruses (HERVd) uses a specialized method tailored for complexities of HERV evolution[72]. TSDfinder[68] defragments LINE1 from the RepeatMasker output and reports target site duplications. In many instances, custom methods for given repetitive classes may be required. Yet it is obvious that any attempt to design a perfect defragmentation method is unrealistic. In fact, some defragmentation schemes may be misleading, and those interested in biological analysis of repetitive DNA should always compare defragmented results with the original ones.

7.6 Applications

7.6.1 Masking and detection of repetitive sequences

Masking repetitive elements is the first step in many analyses of nucleic acids sequences including but not limited to similarity searches, gene prediction, mRNA annotation, and coding sequences (CDS) prediction. For short sequences one can basically use any program performing similarity-based detection of repetitive DNA, provided that the corresponding repetitive libraries are available. See Figures 7-3 and 7-4 for Censor and RepeatMasker web submission forms.

For large-scale or frequent analysis of repetitive elements, it is best to use local installations of these programs. Both programs can filter interspersed and most common classes of tandem repeats. For large sequences, if time is of importance, Censor or MaskerAid based on a fast Wublast search are recommended.

If the goal is to exclusively mask simple repeats, Censor, RepeatMasker or MaskerAid can do the task (look for program manuals how to limit the analysis to simple repeats only). For large sequences it is faster to apply specialized SSR programs that can detect and mask long and/or rare SSR sequences in addition to the common ones, such as Tandem Repeats Finder.

Frequently, repetitive elements themselves are the main subject of analysis. For example, one can be interested in the distribution of repeats in a particular genome. If the query sequences come from a well-studied genome for which Repbase libraries are available, the best and easiest solution is to apply one of the similarity-based detection programs. Censor and RepeatMasker, including RepeatMasker with MaskerAid, produce output files containing precise coordinates of repetitive elements which can be directly used in repeat distribution studies. All the three similarity-based programs can be recommended but, pending specific situations, some are more suitable than others. Comparison of outputs from different programs is always recommended. For large genomes, with high repeat content, the analysis can be very time consuming. One option is to run the programs on multiprocessor or parallel machines. If the time or computer power are limited, a Wublast based search used by Censor or MaskerAid is at least several times faster than RepeatMasker running default Cross_match. However, if sensitivity is the main concern, RepeatMasker with the sensitive settings (-s option) is the most sensitive method so far for nucleic acid comparisons. Nevertheless, if the elements of interest encode proteins, Censor using translated BLAST searches may be most useful as it is able to detect very distant protein similarities undetectable at the nucleic acids level, i.e. query protein vs. translated DNA or translated query DNA vs. translated DNA.

For studies of nucleotide substitution patterns in transposable elements, pairwise alignments of elements and the corresponding consensus sequences are often required. In such situations Censor or RepeatMasker with Cross_match are preferable, because they allow options for producing alignments between elements and consensus sequences. Due to rapid evolution, all hypermutable sites such as simple sequence repeats or CpG sites should be carefully evaluated.

In repeat distribution studies one is often less interested in proportions of a particular repeat in the genomic sequences than in the actual number of independent insertions. As noted above, vertebrate genomes contain large numbers of old elements fragmented by deletions, recombinations or insertions of other elements. Counting such fragments as independently inserted elements would severely bias the results. One remedy is to use RepeatMasker that implements its own defragmentation algorithm. An alternative automatic concatenation method is under development in Censor and will be included in future versions [Kohany et al., in unpublished data]. However, given the complexity of TEs and genomic rearrangements, custom methods may be necessary in more complicated cases.

7.6.2 *De novo* detection of repetitive elements

Detection of transposable elements (TEs) in newly sequenced genomes is a challenging task. The same classes of TEs are often highly divergent between taxonomically distant species and show no clear DNA similarity to each other. Therefore, application of similarity-based programs with libraries from closely related species is limited. In these situations *de novo* identification methods that detect most frequent sequences represent in the genome are recommended. Despite the fact that consensus sequences from such repeats are often incomplete and lack relevant biological information, such knowledge is usually sufficient for genome assembly and masking needed for functional annotation.

There are several drawbacks of *de novo* detection. First of all, one needs to use some thresholds to separate low copy TEs from genome duplications. Whereas in human and vertebrate genomes TEs come from high copy families with hundreds or many thousands copies, the situation in other genomes can be very different. In many insect or plant genomes interspersed repeats are present in just a few copies, or even a single one. Based on pairwise genome cross-comparisons, it is impossible to distinguish such repeats from short duplications or even detect them, if only a single copy is present. For example, the fruit fly (D. melanogaster) genome harbors dozens of different low copy TEs with potentially intact ORFs[73]. Incomplete masking of such repeats can cause contamination of gene prediction by TE-derived sequences. In such situations, it is best to combine *de novo* detection of TEs with protein similarity search, either using Censor with protein sequences from RU, or any translated search with protein sequences from well established protein databases like Swissprot or Genbank, since they contain basic proteins encoded by TEs. *De novo* search can detect high copy repetitive elements, whereas protein search will target autonomous elements. However, further analyses of insertions within TEs and analyses of sequence motifs are necessary for detection of non-coding nonautonomous elements.

Analysis of insertions into known TEs is a very efficient method for identification of new elements. Presence of similar sequences at independent locations is a good indicator that these sequences represent a new transposon family (except for frequent processed pseudogenes, but those are rare). Analysis of target sites is one of the best ways to estimate the nature of a transposable elements (see section on Interspersed repeats). Non-LTR-retrotransposons are flanked by TSDs of variable length (1-20 bp). LTR-elements often contain 5'-TG and CA-3' terminal dinucleotides and generate strict short direct repeats (4 or 5 or 6 bp). DNA transposons are characterized by 2-10 bp TSDs. Also some TEs use particular insertion motifs providing further biological information. Taking together all pieces of evidence such as target sites, presence or absence of polyA and long terminal repeats or inverted repeats, it is possible to accurately classify majority of new elements. This includes nonautonomous TEs which may otherwise be difficult to identify by sequence similarity due to the lack of protein coding sequences.

One exception is nonautonomous rolling circle transposons, because they lack inverted repeats (TIRs) and TSDs. Although there is a 16- to 20-bp palindrome just upstream of the conserved 3' CTRR, preservation of palindrome structures but not the sequence would apparently preclude the applicability of a consensus sequence in the identification of rolling circle DNA transposons by computer-assisted searches [43]. Nevertheless, common 5'-TC and CTRR-3' termini and TA-target specificity may be critical for identification of these elements. CR1-mobilized elements also don't produce TSDs upon insertion. However, autonomous CR1 LINEs can be identified by protein similarity. Although SINEs don't code for any protein, their 3' ends, including CR1-derived SINEs, are frequently derived from 3'ends of corresponding LINE counterparts[32]. Therefore, we can detect them even without analysis of insertion, with the exception of partial sequences.

7.6.3 Gene, CDS prediction and contamination of gene and protein databases with repetitive DNA

The possible contribution of repetitive DNA to coding capacity of the host genome remains a highly controversial issue. There are a very few well confirmed contributions of TE to the human proteom. However, the majority of cases found in databanks are predicted hypothetical proteins or splice variants. For example, frequent detection of noncoding human Alu elements in proteins raised many critical alerts published about gene and CDS prediction and annotation[74-77]. Despite a long effort to detect TEs in human coding DNA, the majority of known cases in databanks are predicted genes or CDS contaminated by TEs. Detailed inspection of a set of 781 well-defined proteins confirmed on the protein level by direct amino acids sequencing or 3D structure determination, completely lack any Alu or other TEs[77].

While preliminary masking of repetitive DNA is a standard procedure in gene prediction for genomic DNA, published annotations of EST or full length cDNA sequences often imply that all detected RNA codes for a protein and simple ORF detection methods are applied without preliminary masking of repetitive DNA. However, the majority of mature RNA polymerase II transcripts represent RNAs without any coding capacity (hn RNA or noncoding RNA genes)[78]. Extraction and reverse transcription of total cellular RNA extracts can result in many unprocessed or defective noncoding RNA in cDNA libraries. While such defective or alternatively spliced RNAs are likely to be regulated or destroyed by cellular mRNA quality control mechanisms, the corresponding CDS and protein sequences are frequently predicted and subsequently found in public databases including curated ones such as Refseq or Swissprot. Many recent high-throughput sequencing cDNA projects are the main source of contamination of protein databases with unconfirmed and TE-containing sequences[Pavlicek et al., unpublished data].

Interspersed repeats contain promoters and many regulatory elements. A contribution of TEs to regulation of cellular transcripts is a controversial issue. Analysts of transcriptional regulators in eukaryotic genomes should be aware of TE-derived sequences and carefully evaluate their potential impact. Cellular genes were apparently well regulated long before insertion of TEs and presence of TE-derived motifs does not seem to be essential for them. Nevertheless, some TEs may be involved in altering gene regulation for specific evolutionary adaptations.

In summary, given the uncertainty regarding real contribution of repetitive elements to host proteom and to regulation of cellular genes, preliminary masking of repetitive DNA should precede not only gene prediction from genomic DNA, but also all cDNA-based CDS prediction and annotation of regulatory elements. Given the relatively high contamination of public databases with inaccurate gene and CDS prediction, we highly recommend masking of query sequences not only for noncoding sequences, but also for coding sequences.

Acknowledgments

We thank Jolanta Walichiewicz and Matthew Jurka for help with editing of this manuscript. This work was supported in part by the grant 2P41LM6252-04A1 from the National Institute of Health.

References

1. Lander, E.S., Linton, L.M., Birren, B., Nusbaum, C., Zody, M.C., Baldwin J., Devon, K., Dewar, K., Doyle, M., FitzHugh, W., Funke, R., Gage, D., Harris, K., Heaford, A., Howland, J., Kann, L., Lehoczky, J., LeVine, R., McEwan, P., McKernan, K., Meldrim, J., Mesirov, J.P., Miranda, C., Morris, W., Naylor, J., Raymond, C., Rosetti, M., Santos, R., Sheridan, A., Sougnez, C., Stange-Thomann, N., Stojanovic, N., Subramanian, A., Wyman, D., Rogers, J., Sulston, J., Ainscough, R., Beck, S., Bentley, D., Burton, J., Clee, C., Carter, N., Coulson, A., Deadman, R., Deloukas, P., Dunham, A., Dunham, I., Durbin, R., French, L., Grafham, D., Gregory, S., Hubbard, T., Humphray, S., Hunt, A., Jones, M., Lloyd, C., McMurray, A., Matthews, L., Mercer, S., Milne, S., Mullikin, J.C., Mungall, A., Plumb, R., Ross, M., Shownkeen, R., Sims, S., Waterston, R.H., Wilson, R.K., Hillier, L.W., McPherson, J.D., Marra, M.A., Mardis, E.R., Fulton, L.A., Chinwalla, A.T., Pepin, K.H., Gish, W.R., Chissoe, S.L., Wendl, M.C., Delehaunty, K.D., Miner, T.L., Delehaunty, A., Kramer, J.B., Cook, L.L., Fulton, R.S., Johnson, D.L., Minx, P.J., Clifton, S.W., Hawkins, T., Branscomb, E., Predki, P., Richardson, P., Wenning, S., Slezak, T., Doggett, N., Cheng, J.F., Olsen, A., Lucas, S., Elkin, C., Uberbacher, E., Frazier, M., Gibbs, R.A., Muzny, D.M., Scherer, S.E., Bouck, J.B., Sodergren, E.J., Worley, K.C., Rives, C.M., Gorrell, J.H., Metzker, M.L., Naylor, S.L., Kucherlapati, R.S., Nelson, D.L., Weinstock, G.M., Sakaki, Y., Fujiyama, A., Hattori, M., Yada, T., Toyoda, A., Itoh, T., Kawagoe, C., Watanabe, H., Totoki, Y., Taylor, T., Weissenbach, J., Heilig, R., Saurin, W., Artiguenave, F., Brottier, P., Bruls, T., Pelletier, E., Robert, C., Wincker, P., Smith, D.R., Doucette-Stamm, L., Rubenfield, M., Weinstock, K., Lee, H.M., Dubois, J., Rosenthal, A., Platzer, M., Nyakatura, G., Taudien, S., Rump, A., Yang, H., Yu, J., Wang, J., Huang, G., Gu, J., Hood, L., Rowen, L., Madan, A., Qin, S., Davis, R.W., Federspiel, N.A., Abola, A.P., Proctor, M.J., Myers, R.M., Schmutz, J., Dickson, M., Grimwood, J., Cox, D.R., Olson, M.V., Kaul, R., Raymond, C., Shimizu, N., Kawasaki, K., Minoshima, S., Evans, G.A., Athanasiou, M., Schultz, R., Roe, B.A., Chen, F., Pan, H., Ramser, J., Lehrach, H., Reinhardt, R., McCombie, W.R., de la Bastide, M., Dedhia, N., Blocker, H., Hornischer, K., Nordsiek, G., Agarwala, R., Aravind, L., Bailey, J.A., Bateman, A., Batzoglou, S., Birney, E., Bork, P., Brown, D.G., Burge, C.B., Cerutti, L., Chen, H.C., Church, D., Clamp, M., Copley, R.R., Doerks, T., Eddy, S.R., Eichler, E.E., Furey, T.S., Galagan, J., Gilbert, J.G., Harmon, C., Hayashizaki, Y., Haussler, D., Hermjakob, H., Hokamp, K., Jang, W., Johnson, L.S., Jones, T.A., Kasif, S., Kaspryzk, A., Kennedy, S., Kent, W.J., Kitts, P., Koonin, E.V., Korf, I., Kulp, D., Lancet, D., Lowe, T.M., McLysaght, A., Mikkelsen, T., Moran, J.V., Mulder, N., Pollara, V.J., Ponting, C.P., Schuler, G., Schultz, J., Slater, G., Smit, A.F., Stupka, E., Szustakowski, J., Thierry-Mieg, D., Thierry-Mieg, J., Wagner, L., Wallis, J., Wheeler, R., Williams, A., Wolf, Y.I., Wolfe, K.H., Yang, S.P., Yeh, R.F., Collins, F., Guyer, M.S., Peterson, J., Felsenfeld, A., Wetterstrand, K.A., Patrinos, A., Morgan, M.J., Szustakowki, J., de Jong, P., Catanese, J.J., Osoegawa, K., Shizuya, H., Choi, S., Chen, Y.J.: Initial sequencing and analysis

of the human genome. *Nature* 2001;409:860-921.

2. Weber, J.L., May, P.E.: Abundant class of human DNA polymorphisms which can be typed using the polymerase chain reaction. *Am. J. Hum. Genet.* 1989;44:388-396.

3. Schlotterer, C., Tautz, D.: Slippage synthesis of simple sequence DNA. *Nucl. Acids Res.* 1992;20:211-215.

4. Jeffreys, A.J., MacLeod, A., Tamaki, K., Neil, D.L., Monckton, D.G.: Minisatellite repeat coding as a digital approach to DNA typing. *Nature* 1991;354:204-209.

5. Jeffreys, A.J., Tamaki, K., MacLeod, A., Monckton, D.G., Neil, D.L., Armour, J.A.: Complex gene conversion events in germline mutation at human minisatellites. *Nat. Genet.* 1994;6:136-145.

6. Pâques, F., Leung, W.Y., Haber, J.E.: Expansions and contractions in a tandem repeat induced by double-strand break repair. *Mol. Cell. Biol.* 1998;18:2045-2054.

7. Richard, G.F., Pâques, F.: Mini- and microsatellite expansions: the recombination connection. *EMBO Rep.* 2000;1:122-126.

8. Vergnaud, G., Denoeud, F.: Minisatellites: mutability and genome architecture. *Genome Res.* 2000;10:899-907.

9. Kovtun, I.V., McMurray, C.T.: Trinucleotide expansion in haploid germ cells by gap repair. *Nat. Genet.* 2001;27:407-411.

10. Li, Y.C., Korol, A.B., Fahima, T., Beiles, A., Nevo, E.: Microsatellites: genomic distribution, putative functions and mutational mechanisms: a review. *Mol. Ecol.* 2002;111:2453-2465.

11. Metzgar, D., Bytof, J., Wills, C.: Selection against frameshift mutations limits microsatellite expansion in coding DNA. *Genome Res.* 2000;10:72-80.

12. Toth, G., Gaspari, Z., Jurka, J.: Microsatellites in different eukaryotic genomes: survey and analysis. *Genome Res.* 2000;10:967-981.

13. Morgante, M., Hanafey, M., Powell, W.: Microsatellites are preferentially associated with nonrepetitive DNA in plant genomes. *Nat. Genet.* 2002;30:194-200.

14. Katti, M.V., Ranjekar, P.K., Gupta, V.S.: Differential distribution of simple sequence repeats in eukaryotic genome sequences. *Mol. Biol. Evol.* 2001;18:1161-1167.

15. Xiong, Y., Eickbush, T.H.: Origin and evolution of retroelements based upon their reverse transcriptase sequences. *EMBO J.* 1990;9:3353-3362.

16. Malik, H.S., Burke, W.D., Eickbush, T.H.: The age and evolution of non-LTR retrotransposable elements. *Mol. Biol. Evol.* 1999;16: 793-805.

17. Malik, H.S., Eickbush, T.H.: Phylogenetic analysis of ribonuclease H domains suggests a late, chimeric origin of LTR retrotransposable elements and retroviruses. *Genome Res.* 2001;11:1187-1197.

18. Becker, K.G., Swergold, G.D., Ozato, K., Thayer, R.E.: Binding of the ubiquitous nuclear transcription factor YY1 to a cis regulatory sequence in the human LINE-1 transposable element. *Hum. Mol. Genet.* 1993;2:1697-1702.

19. Kurose, K., Hata, K., Hattori, M., Sakaki, Y.: RNA polymerase III dependence of the human L1 promoter and possible participation of the RNA polymerase II factor YY1 in the RNA polymerase III transcription system. *Nucleic Acids Res.* 1995;23:3704-3709.

20. Mathias, S.L., Scott, A.F., Kazazian, H.H. Jr., Boeke, J.D., Gabriel, A.: Reverse transcriptase encoded by a human transposable element. *Science* 991;254:1808-1810.

21. Feng, Q., Moran, J.V., Kazazian, H.H. Jr., Boeke, J.D.: Human L1 retrotransposon encodes a conserved endonuclease required for retrotransposition. *Cell* 1996;87:905-916.

22. Hohjoh, H., Singer, M.F.: Sequence-specific single-strand RNA binding protein encoded by the human LINE-1 retrotransposon. *EMBO J.* 1997;16:6034-6043.

23. Martin, S.L., Bushman, F.D: Nucleic acid chaperone activity of the ORF1 protein from the mouse LINE-1 retrotransposon. *Mol. Cell Biol.* 2001;21:467-475.

24. Kapitonov, V.V., Jurka, J.: The Esterase and PHD Domains in CR1-Like Non-LTR Retrotransposons. *Mol. Biol. Evol.* 2003;20:38-46.

25. Jurka, J.: Sequence patterns indicate an enzymatic involvement in integration of mammalian retroposons. *Proc. Natl. Acad. Sci. U.S.A.* 1997;94:1872-1877.

26. Gilbert, N., Lutz-Prigge, S., Moran, J.V.: Genomic deletions created upon LINE-1 retrotransposition. *Cell* 2002;110:315-325.

27. Tatout, C., Lavie, L., Deragon, J.M.: Similar target site selection occurs in integration of plant and mammalian retroposons. *J. Mol. Evol.* 1998;47:463-470.

28. Sasaki, T., Fujiwara, H.: Detection and distribution patterns of telomerase activity in insects. *Eur. J. Biochem.* 2000;267:3025-3031.

29. Kapitonov, V.V., Jurka, J.: A Novel Class of SINE Elements Derived from 5S rRNA. *Mol. Biol. Evol.* 2003;20:694-702.

30. Schmid, C.W.: Does SINE evolution preclude Alu function? *Nucleic Acids Res.* 1998;26:4541-4550.

31. Kajikawa, M., Okada, N.: LINEs mobilize SINEs in the eel through a shared 3' sequence. *Cell* 2002;111:433-444.

32. Okada, N., Hamada, M., Ogiwara, I., Ohshima, K.: SINEs and LINEs share common 3' sequences: a review. *Gene* 1997;205:229-243.

33. Jurka, J.: Repeats in genomic DNA: mining and meaning. *Curr. Opin. Struct. Biol.* 1998;8:333-337.

34. Jurka, J., Kapitonov, V.V.: Sectorial mutagenesis by transposable elements. *Genetica* 1999;107:239-248.

35. Smit, A.F., Riggs, A.D.: MIRs are classic, tRNA-derived SINEs that amplified before the mammalian radiation. *Nucleic Acids Res.* 1995;23:98-102.

36. Repbase Update: http://www.girinst.org/Repbase_Update.html

37. Ullu, E., Tschudi, C.: Alu sequences are processed 7SL RNA genes. *Nature* 1984;312:171-172.

38. Wilkinson, D.: Human endogenous retroviruses; in Levy JA (ed): The Retroviridae. Plenum Press New York, NY, 1994, vol 3, pp 465-553.

39. Smit, A.F.: Interspersed repeats and other mementos of transposable elements in

mammalian genomes. *Curr. Opin. Genet. Dev.* 1999;9:657-663.

40. Sverdlov, E.D.: Retroviruses and primate evolution. *BioEssays* 2000;22:161-171.
41. Smit, A.F.: Identification of a new, abundant superfamily of mammalian LTR-transposons. *Nucleic Acids Res.* 1993;21:1863-1872.
42. Yang, J., Bogerd, H.P., Peng, S., Wiegand, H., Truant, R., Cullen, B.R.: An ancient family of human endogenous retroviruses encodes a functional homolog of the HIV-1 Rev protein. *Proc. Natl. Acad. Sci. U.S.A.* 1999;96:13404-13408.
43. Kapitonov, V.V., Jurka, J.: Rolling-circle transposons in eukaryotes. *Proc. Natl. Acad. Sci. U.S.A.* 2001;98:8714-8719.
44. Jurka, J., Walichiewicz, J., Milosavljevic, A.: Prototypic sequences for human repetitive DNA. *J. Mol. Evol.* 1992;35:286-291.
45. Jurka, J.: Repbase update: a database and an electronic journal of repetitive elements. *Trends Genet.* 2000;16:418-420.
46. Waterston, R.H., Lindblad-Toh, K., Birney, E., Rogers, J., Abril, J.F., Agarwal, P., Agarwala, R., Ainscough, R., Alexandersson, M., An, P., Antonarakis, S.E., Attwood, J., Baertsch, R., Bailey, J., Barlow, K., Beck, S., Berry, E., Birren, B., Bloom, T., Bork, P., Botcherby, M., Bray, N., Brent, M.R., Brown, D.G., Brown, S.D., Bult, C., Burton, J., Butler, J., Campbell, R.D., Carninci, P., Cawley, S., Chiaromonte, F., Chinwalla, A.T., Church, D.M., Clamp, M., Clee, C., Collins, F.S., Cook, L.L., Copley, R.R., Coulson, A., Couronne, O., Cuff, J., Curwen, V., Cutts, T., Daly, M., David, R., Davies, J., Delehaunty, K.D., Deri, J., Dermitzakis, E.T., Dewey, C., Dickens, N.J., Diekhans, M., Dodge, S., Dubchak, I., Dunn, D.M., Eddy, S.R., Elnitski, L., Emes, R.D., Eswara, P., Eyras, E., Felsenfeld, A., Fewell, G.A., Flicek, P., Foley, K., Frankel, W.N., Fulton, L.A., Fulton, R.S., Furey, T.S., Gage, D., Gibbs, R.A., Glusman, G., Gnerre, S., Goldman, N., Goodstadt, L., Grafham, D., Graves, T.A., Green, E.D., Gregory, S., Guigo, R., Guyer, M., Hardison, R.C., Haussler, D., Hayashizaki, Y., Hillier, L.W., Hinrichs, A., Hlavina, W., Holzer, T., Hsu, F., Hua, A., Hubbard, T., Hunt, A., Jackson, I., Jaffe, D.B., Johnson, L.S., Jones, M., Jones, T.A., Joy, A., Kamal, M., Karlsson, E.K., Karolchik, D., Kasprzyk, A., Kawai, J., Keibler, E., Kells, C., Kent, W.J., Kirby, A., Kolbe, D.L., Korf, I., Kucherlapati, R.S., Kulbokas, E.J., Kulp, D., Landers, T., Leger, J.P., Leonard, S., Letunic, I., Levine, R., Li, J., Li, M., Lloyd, C., Lucas, S., Ma, B., Maglott, D.R., Mardis, E.R., Matthews, L., Mauceli, E., Mayer, J.H., McCarthy, M., McCombie, W.R., McLaren, S., McLay, K., McPherson, J.D., Meldrim, J., Meredith, B., Mesirov, J.P., Miller, W., Miner, T.L., Mongin, E., Montgomery, K.T., Morgan, M., Mott, R., Mullikin, J.C., Muzny, D.M., Nash, W.E., Nelson, J.O., Nhan, M.N., Nicol, R., Ning, Z., Nusbaum, C., O'Connor, M.J., Okazaki, Y., Oliver, K., Overton-Larty, E., Pachter, L., Parra, G., Pepin, K.H., Peterson, J., Pevzner, P., Plumb, R., Pohl, C.S., Poliakov, A., Ponce, T.C., Ponting, C.P., Potter, S., Quail, M., Reymond, A., Roe, B.A., Roskin, K.M., Rubin, E.M., Rust, A.G., Santos, R., Sapojnikov, V., Schultz, B., Schultz, J., Schwartz, M.S., Schwartz, S., Scott, C., Seaman, S., Searle, S., Sharpe, T., Sheridan, A., Shownkeen, R., Sims, S., Singer, J.B., Slater, G., Smit, A., Smith, D.R., Spencer, B., Stabenau, A., Stange-Thomann, N., Sugnet, C., Suyama, M.,

Tesler, G., Thompson, J., Torrents, D., Trevaskis, E., Tromp, J., Ucla, C., Ureta-Vidal, A., Vinson, J.P., Von Niederhausern, A.C., Wade, C.M., Wall, M., Weber, R.J., Weiss, R.B., Wendl, M.C., West, A.P., Wetterstrand, K., Wheeler, R., Whelan, S., Wierzbowski, J., Willey, D., Williams, S., Wilson, R.K., Winter, E., Worley, K.C., Wyman, D., Yang, S., Yang, S.P., Zdobnov, E.M., Zody, M.C., Lander, E.S.: Initial sequencing and comparative analysis of the mouse genome. *Nature* 2002;420:520-562.

47. Jurka, J., Pethiyagoda, C.: Simple repetitive DNA sequences from primates: compilation and analysis. *J. Mol. Evol.* 1995;40:120-126.

48. Milosavljevic, A., Jurka, J.: Discovering simple DNA sequences by the algorithmic significance method. *Comput. Appl. Biosci.* 1993;9:407-411.

49. Benson, G.: Sequence alignment with tandem duplication. *J. Comput. Biol.* 1997;4:351-367.

50. Benson, G.: Tandem repeats finder: a program to analyze DNA sequences. *Nucleic Acids Res.* 1999;27:573-580.

51. Rice, P., Longden, I., Bleasby, A.: EMBOSS: The European Molecular Biology Open Software Suite. *Trends Genet.* 2000;16:276-277.

52. Bao, Z., Eddy, S.R.: Automated de novo identification of repeat sequence families in sequenced genomes. *Genome Res.* 2002;12:1269-1276.

53. McCarthy, E.M., McDonald, J.F.: LTR_STRUC: a novel search and identification program for LTR retrotransposons. *Bioinformatics* 2003;19:362-367.

54. Gish, W.: Wublast http://blast.wustl.edu.

55. Green, P.: Phrap http://www.phrap.org/.

56. Jurka, J., Klonowski, P., Dagman, V., Pelton, P.: CENSOR—a program for identification and elimination of repetitive elements from DNA sequences. *Comput. Chem.* 1996;20:119-121.

57. Smith, T.F., Waterman, M.S.: Identification of common molecular subsequences. *J. Mol. Biol.* 1981;147:195-197.

58. Bedell, J.A., Korf, I., Gish, W.: MaskerAid: a performance enhancement to RepeatMasker. *Bioinformatics* 2000;16:1040-1041.

59. Ostertag, E.M., Kazazian, H.H . Jr.: Twin priming: a proposed mechanism for the creation of inversions in L1 retrotransposition. *Genome Res.* 2001;11:2059-2065.

60. Holmes, S.E., Dombroski, B.A., Krebs, C.M., Boehm, C.D., Kazazian, H.H. Jr.: A new retrotransposable human L1 element from the LRE2 locus on chromosome 1q produces a chimaeric insertion. *Nat. Genet.* 1994;7:143-148.

61. Goodier, J.L., Ostertag, E.M., Kazazian, H.H. Jr.: Transduction of 3'-flanking sequences is common in L1 retrotransposition. *Hum. Mol. Genet.* 2000;9:653-657.

62. Pickeral, O.K., Makalowski, W., Boguski, M.S., Boeke, J.D.: Frequent human genomic DNA transduction driven by LINE-1 retrotransposition. *Genome Res.* 2000;10:411-415.

63. Hu WS, Temin HM: Retroviral recombination and reverse transcription. *Science* 1990;250:1227-1233.

64. Goodchild, N.L., Freeman, J.D., Mager, D.L.: Spliced HERV-H endogenous

retroviral sequences in human genomic DNA: evidence for amplification via retrotransposition. *Virology* 1995;206:164-173.
65. Pavlicek, A., Paces, J., Zika, R., Hejnar, J.: Length distribution of long interspersed nucleotide elements (LINEs) and processed pseudogenes of human endogenous retroviruses: implications for retrotransposition and pseudogene detection. *Gene* 2002; 300:189-194.
66. Costas, J.: Characterization of the intragenomic spread of the human endogenous retrovirus family HERV-W. *Mol. Biol. Evol.* 2002;19:526-533.
67. Arco, S.S., Wang, Z., Weber, J.L., Deininger, P.L., Batzer, M.A.: Alu repeats: a source for the genesis of primate microsatellites. *Genomics* 1995;29:136-144.
68. Szak, S.T., Pickeral, O.K., Makalowski, W., Boguski, M.S., Landsman, D., Boeke, J.D.: Molecular archeology of L1 insertions in the human genome. *Genome Biol.* 2002;3:research0052.
69. Jabbari, K., Bernardi, G.: CpG doublets, CpG islands and Alu repeats in long human DNA sequences from different isochore families. *Gene* 1998;224:123-127.
70. Jurka, J., Krnjajic, M., Kapitonov, V.V., Stenger, J.E., Kokhanyy, O.: Active Alu elements are passed primarily through paternal germlines. *Theor. Popul. Biol.* 2002;61:519-530.
71. Mager, D.L., Goodchild, N.L.: Homologous recombination between the LTRs of a human retrovirus-like element causes a 5-kb deletion in two siblings. *Am. J. Hum. Genet.* 1989;45:848-854.
72. Paces, J., Pavlicek, A., Paces, V.: HERVd: database of human endogenous retroviruses. *Nucl. Acids Res.* 2002;30:205-206.
73. Kapitonov, V.V., Jurka, J.: Molecular paleontology of transposable elements in the Drosophila melanogaster genome. *Proc. Natl. Acad. Sci. U.S.A.* in press.
74. Claverie, J.M., Makalowski, W.: Alu alert. *Nature* 1994;371:752.
75. Tugendreich, S., Feng, Q., Kroll, J., Sears, D.D., Boeke, J.D., Hieter, P.: Alu sequences in RMSA-1 protein? *Nature* 1994;370:106.
76. Zietkiewicz, E., Makalowski, W., Mitchell, G.A., Labuda, D.: Phylogenetic analysis of a reported complementary DNA sequence. *Science* 1994;265:1110-1111.
77. Pavlicek, A., Clay, O., Bernardi, G.: Transposable elements encoding functional proteins: pitfalls in unprocessed genomic data? *FEBS Lett* 2002;523:252-253.
78. Jackson, D.A., Pombo, A., Iborra, F.: The balance sheet for transcription: an analysis of nuclear RNA metabolism in mammalian cells. *FASEB J.* 2000;14:242-254.
79. Sagot, M.F., Myers, E.W.: Identifying satellites and periodic repetitions in biological sequences. *J. Comput. Biol.* 1998;5:539-553.

Chapter 8

MatInspector: Analysing Promoters for Transcription Factor Binding Sites

Kerstin Cartharius

Contact: Kerstin Cartharius

Genomatix Software GmbH, Landsberger Str. 6, 80333 München, Germany
E-mail: cartharius@genomatix.de

8.1 Introduction

Major efforts have been made to identify all potential protein-coding sequences within the completely sequenced genomes that are available by now. But also the non-coding sequences that constitute more than 95% of e.g. the human genome play an important part in cell biology. In higher organisms it is vital for the cell that the right gene is activated at the right time during development and differentiation. These complex patterns of gene expression depend on an enormous variety of regulatory units found in the non-coding regions of the genome. Elements like S/MARs, enhancers and promoters all play a role in the regulation of gene transcription.

Promoters are an integral part of the gene, mediating and controlling initiation of transcription. They regulate the expression of the corresponding gene via so-called transcription factor binding sites. Detection of these binding sites is an important task in computer aided promoter analysis.

The computer program *MatInspector* was one of the first to use a weight matrix approach to locate transcription factor binding sites in promoter sequences. It features an extensive precompiled library of binding sites, which made it very popular on the internet. Since the first publication [1] the program itself and the library have evolved considerably. MatInspector's basic algorithm and new improvements will be detailed in this article.

After a putative transcription factor binding site has been located with computer methods the question of its functionality in the promoter arises. This is a topic that can ultimately be answered only by a wet-lab experiment with defined settings, since potential binding sites in a promoter can be functional in certain cells, tissues or development stages and nonessential in other circumstances.

Nevertheless, being able to predict potentially functional binding sites is a first step in in-depth promoter analysis. Especially coregulated promoters often show a clearly defined organization, i.e. common order and distances between the binding sites involved. So the detection of organizational features of binding sites can give further evidence on the mechanisms of regulation and is feasible with the help of computer programs based on MatInspector. The resulting regulatory models can be used to identify additional target genes with similar regulatory properties in genomic sequence databases, leading to a better understanding of the regulatory networks that are active in each cell.

8.2 Promoters and their organization

Control of transcription initiation is a crucial step in gene expression. It is the most important mechanism for determining whether or not a gene is expressed and how much mRNA — and consequently proteins — is produced. A promoter is a sequence that initiates and regulates transcription of a gene. Although there are different levels of regulation in gene expression such as DNA methylation, chromatin structure, or enhancers, studying promoter regions can give detailed information on the regulation of a gene as promoters represent a pivotal part of the expression regulation.

163

Most genes in higher eukaryotes are transcribed from polymerase II (polII) dependent promoters. The transcription process starts after a transcription complex consisting of polII and several general transcription factors such as TBP (TATA binding protein) has been recruited to the promoter (Figure 8-1). In a first step a variety of transcription factors bind to upstream promoter and enhancer sequences to form a multi-protein complex, the activator complex. In a second step this complex recruits the RNA polymerase complex to the correct location on the genomic DNA, i.e. the transcription start site (TSS). It is important to note that both the well-known TATA and CAAT boxes are neither necessary nor sufficient for promoter function in general.

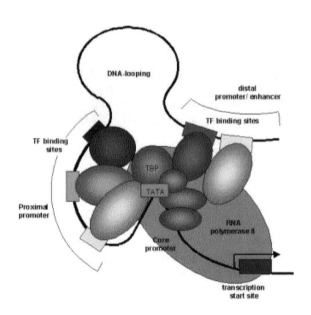

Figure 8-1. *Schematic illustration of transcription factors binding to the promoter, forming the activation complex. Transcription factors are depicted as circles, their binding sites as squares on the DNA sequence.*

The region containing the elements allowing specific initiation of transcription is called the *core promoter*. The *proximal promoter* contains transcription factor binding sites for initiation and fine-tuning of the transcription process and spans a region of up to 1000 basepairs upstream of the transcription start site. *Distal promoter* or *enhancer* sequences can additionally activate or silence transcription of the responsive gene. They often are independent of orientation and position relative to the core promoter region because of DNA looping. Enhancers are also rich in transcription factor binding sites.

Analysis of eukaryotic promoters has shown flexible spacing between transcription factor binding sites within certain limits to have no effect on the promoter function. However, in case of binding sites that interact functionally on short distance alterations of the spacing is tolerated only within very small limits of a few basepairs. This can be explained by the flexible protein domains of transcription factors: when the DNA binding domain of neighbouring transcription factors are shifted slightly, their activation domains may still be able to interact because they are attached to their DNA binding domains by flexible protein regions. However, the protein chains cannot accommodate longer distance variation. In case of longer spacers the interaction of binding proteins is possible via different mechanisms such as DNA looping or proteins that do not directly bind the DNA but can bridge between two DNA binding proteins (so-called mediators of transcription).

8.3 Transcription Factors and their Binding Sites

Although other promoter elements like intrinsically curved DNA, direct or inverted repeats may also contribute to promoter function, protein binding sites represent the most crucial elements. The corresponding proteins are called *transcription factors* (TFs). There is a large variety of TFs in the cell, currently more than 1000 are known and a total of about 3000 was estimated for the human genome[2]. TFs usually exhibit two functional domains: a DNA binding domain and one or more activation domains, which interact with other proteins to activate or repress the transcription of the promoter. One important function of a TF is to bring its activation domain into a specific location by binding to the genomic DNA at selected binding sites or cis-regulatory sequences. The influence of this binding on the transcription of a gene is usually determined by the functional context, i.e. which other factors are close enough to interact with the particular factor.

TF binding sites are generally short stretches of DNA, about 8-20 nucleotides in length. It is important to note that TF binding sites only carry the potential to bind their corresponding protein. However, they occur everywhere in the genome and are by no means restricted to regulatory regions. Sites outside regulatory regions are known to bind their transcription factors[3], and it is the context that differentiates a functional binding site affecting gene regulation from a mere physical binding site.

When comparing several binding sites for the same protein, the sequences may vary considerably at different positions and sequence variations are not evenly distributed over the binding site. On the one hand there are highly conserved nucleotide positions that are rarely exchanged. Experiments have shown that the binding ability declines considerably if mutations are introduced at these positions (e.g. [4]). On the other hand there are positions with low or no sequence conservation. This can be explained by the binding mechanism of TFs: the conserved positions represent nucleotides that are contacted by the protein in a sequence-specific manner, whereas the non-conserved positions are internal spacers that may contribute to backbone contacts to the DNA[5]. In order to design an optimal algorithm for detecting binding sites it is important to know the molecular basis of this flexibility.

8.4 *In silico* detection of transcription factor binding sites with MatInspector

The first step in the process of defining a pattern for TF binding sites (TFBSs) is the collection of known binding sites for each TF, the so-called training set. The first idea would be to use a IUPAC consensus to represent the search pattern by aligning and transforming the alignment into a single description. The IUPAC ambiguity code is an extended alphabet of 15 letters which includes symbols for alternative nucleotides (e.g. W = A or T). Assigning a IUPAC code to each position of the alignment is usually done by choosing a representative majority for one or more nucleotides (formalized e.g. by [6]), thus neglecting the original frequency information. A search algorithm based on simple string matching can then be applied to find new locations of the consensus pattern in DNA sequences.

This simple method has three major limitations: First the IUPAC approach cannot detect sequences that contain nucleotides that were not present in any of the training sequences at a particular position. For example, if the training set only contained A or G or T at a certain position, a match with a C would be rejected even if all other positions were perfect matches. Secondly, IUPAC consensus sequences also accept nucleotide sequences that are far off real binding sites but are combinatorial possibilities of the IUPAC consensus composition. The third limitation is the qualitative nature of a IUPAC consensus match: There is only a YES/NO decision but quantitative aspects of how similar a sequence is to a real binding site are neglected.

A better way to detect TFBSs is a nucleotide weight matrix (NWM). The complete composition of nucleotides for each position is used to search a matching sequence and also to quantitatively score the match. The concept of NWMs was developed in the 1980s, but the widespread use of the concept in form of programs was delayed almost a decade since only a few special matrices had been defined. *MatInspector* was one of the first programs to close this gap in 1995, offering an extensive precompiled library of TF binding site matrices (then based on the TRANSFAC database [7]).

8.5 Definition of a weight matrix

A collection of 5 to 10 different, experimentally verified TFBS is required to construct a nucleotide weight matrix for the recognition of a TFBS. This size of training set will allow representing the diversity of a site sufficiently for practical purposes. In a first step the sequences are aligned and counting the bases at each position creates a nucleotide distribution matrix.

To model the binding mechanism of transcription factors with conserved positions and internal spacers within the binding site, MatInspector introduced an efficient and sensitive position weighting: As mentioned before, the strength in conservation of a certain position within a binding site is relevant to its function. Higher conserved positions in the DNA have higher impact on the binding strength and the specificity with respect to the transcription factor. To reflect such a profile of a binding site, a nucleotide distribution matrix is complemented with the so-called *Ci-vector*, constructed for each position i as follows:

(1) $Ci(i) = (100 / \ln5) * (\Sigma (P(i,b) * \ln P(i,b)) + \ln5)$

where

- $b = \{A,C,G,T,gap\}$
- $P(i,b)$ is the relative frequency of nucleotide b at position i
- $0 = Ci = 100$

This Ci-vector represents the conservation of the individual nucleotide positions in the matrix in numerical values and is later used by the search program MatInspector. This formula is based on the entropy definition by Shannon, which was adapted for nucleotide sequences[8] and adjusted to include gaps. Additionally a normalizing step was introduced so that a Ci-value of 100 represents a position with total conservation of one nucleotide, whereas a position with equal distribution of all four nucleotides and gaps shows a Ci-value of 0.

The matrix generation program also determines a core region within the matrix, which is defined by the four consecutive nucleotide positions with the maximum sum of Ci-values. This core region of the matrix is later used by MatInspector to preselect potential matches but has no influence on the matrix score. In the original algorithm the length of the matrix is determined automatically by cutting off low-conserved positions at both matrix ends.

The latest version of the algorithm (MatInspector professional; Genomatix Software) uses only sequences for matrix generation that are recognized by the final matrix: single training sequences that do not fit to the others (i.e. they have a low matrix similarity, see formula (3) below) are rejected.

The matrices in the MatInspector library are derived from single publications with either a nucleotide distribution matrix or a list of binding sites or from several papers where individual binding sites were published. It should be noted, that in contrast to other methods the library is not supposed to represent all literature regarding TFBS but rather the best of current knowledge in terms of specificity and sensitivity. Thus, erroneous binding sites are not represented in a matrix and the strand orientation for some binding sites might be inverted as to respective literature. Low quality matrices are automatically removed from the MatInspector library. The current library contains 634 matrices and is developed independently from the TRANSFAC database. Today only less than 50% of the library still have links to TRANSFAC in contrast to the original library from 1995

167

The following example will illustrate the definition of a matrix for the MatInspector library: eight binding sites for the transcription factor SREBP (sterol regulatory element binding protein) were selected from various papers for the creation of a matrix description. The following table shows the alignment of the sequences with the core marked in bold.

Sequence name	Sequence (core sequence in bold)	Matrix similarity
HMG_1	GTC**TCAC**CCCACTTC	0.963
FPP	TAC**TCAC**ACGAGGTC	0.835
LDL	AAA**TCAC**CCACTGC	0.990
Squalence	TTA**TCAC**GCCAGTCT	0.941
SREBP-2	CCA**TCAC**CCACGCA	0.983
CYP51_hum	AGA**TCAC**CTCAGGCG	0.973
CYP51_rat	AGA**TCAC**CTCAGCAG	0.964
HMG_3	CGG**GCAC**CGCACCAT	0.838

Table 8-1. Training sequences for creation of the SREBP matrix.

The calculation of the Ci-value for the first position of the core (fourth position of the matrix) will be detailed as an example:

$Ci(4) = (100 / \ln 5) *$
$((\quad P(4,A) * \ln P(4,A) +$
$P(4,C) * \ln P(4,C) +$
$P(4,G) * \ln P(4,G) +$
$P(4,T) * \ln P(4,T)) + \ln 5) =$
$(100 / \ln 5) * (0 + 0 + 1/8 * \ln 1/8 + 7/8 * \ln 7/8) + \ln 5) = 76.59$

The graphics below shows the profile of the Ci-vector, i.e. the Ci-value for each position of the matrix description:

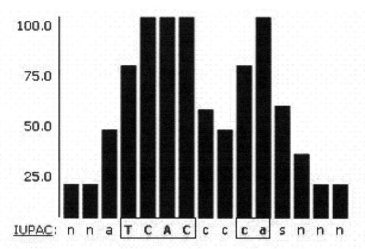

Figure 8-2. Ci-vector profile and IUPAC representation for the SREBP matrix. Basepairs marked by a box show a high information content, i.e. the matrix exhibits a high conservation (Ci-value > 60) at this position. Basepairs in capital letters denote the core sequence.

The frequency of matrix matches in a sequence can be estimated by a value for random expectation termed re-value, which is defined as the number of matches of high quality (matrix similarity $\geq$ 0.85) expected in a random sequence of 1000 basepairs. The calculation of the re-value is a formula that transforms the Ci-value into a probability for each position of the sequence by interpolating between the two endpoints $100 \rightarrow 0.25$ (complete conservation) and $13.9 \rightarrow 1.00$ (random occurrence of the four bases A,C,G and T, no gap). The probabilities at each position are then multiplied to get the re-value for the complete matrix. The re-value for the SREBP matrix above is 0.13 matches per 1000 bps. Only matrices with a re-value below 5 are included in the MatInspector matrix library. More than 70 % of all matrices have a re-value below 1 underlining the high quality of the matrix library.

However, it has to be stressed that the re-value is only a very rough estimate for the number of matches to expect when using a matrix similarity of 0.85. Especially if the *optimized matrix threshold* (see later in this article) is used for searching, the re-value is not suitable to estimate the match number for a matrix.

8.6 Search for matrix matches

The program MatInspector uses the information of core positions, nucleotide distribution matrix and Ci-vector to scan sequences of unlimited length for matches to the matrix description. It should be emphasized that the IUPAC representation for a matrix is not used by the search program but is only given as an illustration for a human reader. The search starts with an optional preselection where only matches to the core region are considered. This reduces the total number of matches and simultaneously accelerates the performance of the program.

First, the core similarity is calculated for each position of the sequence:

(2) core_sim = $(\Sigma\ (\ score(b,j))) / (\Sigma\ (max\_score(j)))$

where
- l is the length of the core region
- j=1..l
- score(b,j): matrix value for base b at position j
- max_score(j): max {score(b,j)} with b in A,C,G,T
- $0 \leq core\_sim \leq 1$

If the core similarity reaches a user defined threshold the matrix similarity is calculated for the corresponding matrix positions:

(3) mat_sim = $(\Sigma\ (Ci(j)*score(b,j)))/(\ \Sigma\ (Ci(j)*max\_score(j)))$

where
- n: length of matrix
- j=1..n
- Ci(j): Ci-value of position j
- score(b,j): matrix-value for base b at position j
- max_score(j): max {score(b,j)} with b = {A,C,G,T }
- $0 \leq mat\_sim \leq 1$

The matrix similarity reaches 1 only if the candidate sequence corresponds to the most conserved nucleotide at each position of the matrix. Multiplying each score with the Ci-value emphasizes the fact that mismatches at less conserved positions are easier tolerated than mismatches at highly conserved positions representing the protein contacts to the sequence. The output of MatInspector consists of those matches that reach a user-defined minimum core and matrix similarity.

Table 8-2 shows the output of MatInspector on 100 basepairs of the promoter sequence for fatty acid synthase in rat (basepairs 1400-1500 of RNFAACSY, accession number X54671). The list includes a binding site for SREBP at position 42 which has been experimentally verified in[9].

Inspecting sequence RNFAACSY_1400 [X54671] (1 - 100): [XXX] [Rat gene for fatty acid synthase (promoter region, exons 1-2).]

Matrix name	from - to	Strand	Core sim.	Matrix sim.	Sequence
IK2.01	12 – 24	(-)	1.000	0.915	agccGGGAccctg
AP2.01	19 – 31	(-)	0.976	0.932	cgCCCCcagccgg
ZBP89.01	22 – 36	(-)	1.000	0.938	cgcgccgCCCCcagc
ZF9.01	22 – 36	(-)	1.000	0.927	cgcgCCGCccccagc
ZF5.01	25 – 35	(+)	0.875	0.906	ggggGCGGcgc
SP1.01	24 – 38	(+)	1.000	0.938	tgggGGCGgcgcgcg
GC.01	24 – 38	(+)	1.000	0.917	tgggGGCGGgcgcgcg
ZF5.01	29 – 39	(-)	1.000	0.963	gcgcGCGCcgc
ZF5.01	30 - 40	(+)	1.000	0.937	cggcGCGCgcg
HES1.01	29 - 43	(+)	0.944	0.936	gcggcgcGCGCgggc
ZF5.01	31 - 41	(-)	1.000	0.937	ccgcGCGCgcc
MYCMAX.03	30 - 44	(-)	0.847	0.905	tgccCGCGcgcgccg
ZF5.01	32 - 42	(+)	1.000	0.966	gcgcGCGCggg
ZF5.01	34 - 44	(+)	0.875	0.921	gcgcGCGGgca
SREBP.02	42 - 56	(+)	1.000	0.988	gcaTCACcccaccga
SREBP.01	42 - 56	(+)	1.000	0.968	gcaTCACcccaccga
ZF5.01	59 - 69	(-)	1.000	0.932	cggcGCGCcgc
ZF5.01	60 - 70	(+)	1.000	0.932	cggcGCGCcgg
IK2.01	67 - 79	(-)	1.000	0.913	ccccGGGAcccgg
ZF5.01	75 - 85	(+)	1.000	0.910	cgggGCGCagc
WHN.01	87 - 97	(+)	1.000	0.965	ccgACGCtcat

Table 8-2. MatInspector output for part of the fatty acid synthase promoter. MatInspector thresholds: matrix similarity=0.9 / core similarity=0.75.

The output shows two general shortcomings of the original MatInspector algorithm: the binding site for SREBP at position 42 is listed twice, as there are several similar SREBP descriptions in the MatInspector library that are based on different publications and experiments. Additionally, the matrix ZF5 shows a high frequency of matches, probably false positive sites. Observations like these led to improvements of the MatInspector library and algorithm as will be detailed below.

8.7 Enhanced MatInspector features

8.7.1 The MatInspector Family concept

Similar matrices for one transcription factor can lead to multiple matches at the same position or matches that are only shifted by a few basepairs if the corresponding matrix descriptions differ in length. It is reasonable to keep different descriptions for a factor in the library as those matrices might be based on different training datasets originating from independent publications. A new feature of the matrix generation algorithm is the assignment of matrix families to each individual matrix. Such a family consists of all matrices that represent similar DNA patterns and are assigned to transcription factors with a similar biological function. The concept of matrix families does not compromise the specificity of the individual matrix approach as a combined matrix including all the individual matrices would do.

MatInspector applies a further step and compares the matches of matrices that belong to the same family. The program only lists the best of a number of overlapping matches of a family in the output. The family concept leads to a significantly reduced output because redundant matches are eliminated. All matrices of a matrix family are of the same (uneven) length resulting in identical positions for the individual matrix matches of a family. The central position of a matrix is called the *anchor* of a family and is listed in the output together with the start and end position of a match.

The following example details the consequences of the introduction of matrix families: The family V$NFKB (Nuclear Factor Kappa B/c-rel) comprises 5 matrices for NFkappaB listed below as well as one matrix for the NFkappaB related factor c-Rel (V$CREL.01).

Family	Matrices	Description	RE-value
V$NFKB	V$NFKAPPAB.01	NF-kappaB	0.42
	V$NFKAPPAB65.01	NF-kappaB (p65)	0.09
	V$CREL.01	c-Rel	2.74
	V$NFKAPPAB50.01	NF-kappaB (p50)	0.05
	V$NFKAPPAB.02	NF-kappaB	0.05
	V$NFKAPPAB.03	NF-kappaB	0.28

Table 8-3. Matrices in the matrix family V$NFKB.

When searching transcription factor binding sites for NFkappaB in a HIV-1 long terminal repeat sequence (HIVTH475A) with the 6 individual matrices 14 matches are found, although only 4 distinct sites are detected:

Inspecting sequence HIVTH475A [L31963] (1 - 500):[RNA] [Human immunodeficiency virus type 1 (individual isolate: TH4-7-5) gene.]

Family/matrix	Position from - to	anchor	Str.	Core sim.	Matrix sim.	Sequence
V$NFKB/NFKAPPAB.01	345 - 359	352	(+)	1.000	0.988	aaGGGActttccgct
V$NFKB/NFKAPPAB65.01	345 - 359	352	(+)	1.000	0.984	aagggactTTCCgct
V$NFKB/CREL.01	345 - 359	352	(+)	1.000	0.983	aagggactTTCCgct
V$NFKB/NFKAPPAB.02	345 - 359	352	(+)	1.000	0.977	aaGGGActttccgct
V$NFKB/NFKAPPAB.03	345 - 359	352	(+)	1.000	0.968	aaGGGActttccgct
V$NFKB/NFKAPPAB.02	359 - 373	366	(+)	1.000	1.000	tgGGGActttccagg
V$NFKB/NFKAPPAB.01	359 - 373	366	(+)	1.000	0.988	tgGGGActtccagg
V$NFKB/NFKAPPAB65.01	359 - 373	366	(+)	1.000	0.984	tggggactTTCCagg
V$NFKB/CREL.01	359 - 373	366	(+)	1.000	0.983	tggggactTTCCagg
V$NFKB/NFKAPPAB.03	359 - 373	366	(+)	1.000	0.994	tgGGGActtccagg
V$NFKB/NFKAPPAB.03	385 - 399	392	(+)	1.000	0.845	tgGGGAactttccgc
V$NFKB/NFKAPPAB.03	386 - 400	393	(+)	0.775	0.893	ggGGAActttccgct
V$NFKB/NFKAPPAB.02	400 - 414	407	(+)	1.000	1.000	tgGGGActttccagg
V$NFKB/NFKAPPAB.01	400 - 414	407	(+)	1.000	0.988	tgGGGActttccagg
V$NFKB/NFKAPPAB65.01	400 - 414	407	(+)	1.000	0.984	tggggactTTCCagg
V$NFKB/CREL.01	400 - 414	407	(+)	1.000	0.983	tggggactTTCCagg
V$NFKB/NFKAPPAB.03	400 - 414	407	(+)	1.000	0.994	tgGGGActttccagg

Table 8-4. Searching NFKB sites with individual matrices.

With matrix families the same search results in exactly 4 matches for the NFkappaB family, which are all biologically significant as described in[10].

Family/matrix	Position from - to	anchor	Str.	Core sim.	Matrix sim.	Sequence
V$NFKB/NFKAPPAB.01	345 - 359	352	(+)	1.000	0.988	aaGGGActttccgct
V$NFKB/NFKAPPAB.02	359 - 373	366	(+)	1.000	1.000	tgGGGActttccagg
V$NFKB/NFKAPPAB.03	386 - 400	393	(+)	0.775	0.893	ggGGAActttccgct
V$NFKB/NFKAPPAB.02	400 - 414	407	(+)	1.000	1.000	tgGGGActttccagg

Table 8-5. Searching NFKB sites with matrix families.

The assignment of matrices from the MatInspector library into families is primarily based on DNA sequence and not on protein classification since a TF can have more than one distinct binding site. For example, the paired domain and the homeo domain binding sites of Pax4/Pax6 are assigned to different families as their respective binding sequences differ considerably. In a first step the matrices are grouped automatically by comparing the similarity of the matrices. For this grouping a neural network approach with self-organizing maps is used. In a second step the resulting groups are checked for biological significance and correctness by evaluating the corresponding literature. If new matrices are added to the library they are mapped onto the existing map of matrices to check whether they are similar to an existing family or if a new family has to be created.

The current library (version 5.0, February 2005) of MatInspector available at www.genomatix.de contains a total of 634 matrices in 279 families, divided into the sections vertebrates, fungi, insects, plants and miscellaneous for transcription factors and a section others, that contains patterns like PolyA signals.

Library Section	Number of Families	Number of Matrices
Vertebrates	150	409
Fungi	32	43
Plants	58	126
Insects	27	40
Miscellaneous	7	8
Others	5	8

Table 8-6. Number of matrix families and matrices in the MatInspector library version 5.0.

8.7.2 Optimized Matrix Thresholds

With the original version of MatInspector it was apparent that with a given matrix similarity threshold some matrices occurred very frequently in almost every sequence whereas others hardly appeared at all, even missing true positive matches in test sequences. The reason is the different length and conservation profile of the matrices in the library: the probability for matches to a long and highly conserved matrix to reach a fixed threshold is low, whereas short, less conserved matrices quickly reach that threshold leading to false positive matches in the output. One indication for a high number of occurrences is a high re-value even though its calculation is based on random sequences. To avoid many false positive matches as a result of less specific matrix descriptions a so-called *optimized matrix threshold* was introduced for each matrix in the library.

The optimized threshold of a weight matrix is defined as the matrix similarity threshold that minimizes false positive matches in the MatInspector output. It is calculated in a way that at most 3 matches are found in 10,000 bps of non-regulatory test sequences, i.e. with the optimized matrix threshold less than 3 false positives per 10,000 bps are found. The latest version of MatInspector (www.genomatix.de) uses the optimized matrix threshold for each matrix as default.

Now, highly specific or relatively long matrices generally have a lower value for the optimized matrix similarity than less specific or shorter matrices. Therefore, an absolute matrix similarity of e.g. 0.85 is a good match to a matrix with an optimized matrix threshold of 0.77 but a poor match to a matrix with 0.93 as optimized matrix threshold.

The following example will detail the consequences of the introduction of optimized matrix thresholds. When searching glucocorticoid response elements (GRE) in a HIV-1 sequence (HIVBRUCG) with a fixed matrix similarity threshold of 0.85 the following four matches are found. However, the functional GRE site in the vif open reading frame identified by[11] is missing.

Inspecting sequence HIVBRUCG [K02013] (1 - 9229): [RNA] [Human immunodeficiency virus type 1, isolate BRU, complete genome (LAV-1).]

Family/matrix	Opt.	Position from - to	anchor	Str.	Core sim.	Matrix sim.	Sequence
V$GREF/PRE.01	0.84	740 - 758	749	(-)	1.000	0.887	ttgcccctggaTGTTctgc
V$GREF/PRE.01	0.84	6865 - 6883	6874	(-)	1.000	0.882	tatttccaaatTGTTctct
V$GREF/PRE.01	0.84	7470 - 7488	7479	(-)	1.000	0.869	cctcagcaaatTGTTctgc
V$GREF/PRE.01	0.84	8866 - 8884	8875	(-)	1.000	0.888	taacaagctggTGTTctct

Table 8-7. Searching HIV1 for GRE with a fixed matrix similarity threshold of 0.85.

The same search with optimized matrix threshold locates seven matches for glucocorticoid response elements since the optimized matrix thresholds for all members of the GRE family are below 0.85. The match list now includes the functional GRE site in the vif open reading frame (position 4995 to 5013):

Inspecting sequence HIVBRUCG [K02013] (1 - 9229): [RNA] [Human immun-odeficiency virus type 1, isolate BRU, complete genome (LAV-1).]

Family/matrix	Opt.	Position from - to	anchor	Str.	Core sim.	Matrix sim.	Sequence
V$GREF/PRE.01	0.84	740 - 758	749	(-)	1.000	0.887	ttgcccctggaTGTTctgc
V$GREF/PRE.01	0.84	3195 - 3213	3204	(-)	0.750	0.842	tgggcacccctCGTTcttg
V$GREF/GRE.01	0.81	4995 - 5013	5004	(-)	0.927	0.825	ggctaactatatGTCCtaa
V$GREF/PRE.01	0.84	6865 - 6883	6874	(-)	1.000	0.882	tatttccaaatTGTTctct
V$GREF/ARE.01	0.80	7197 - 7215	7206	(-)	1.000	0.819	ctcggacccatTGTTgtta
V$GREF/PRE.01	0.84	7470 - 7488	7479	(-)	1.000	0.869	cctcagcaaatTGTTctgc
V$GREF/PRE.01	0.84	8866 - 8884	8875	(-)	1.000	0.888	taacaagctggTGTTctct

Table 8-8. Searching GRE with optimized matrix threshold.

The following table summarizes the effects of the different MatInspector features on the number of matches that are listed for the analysis of 300 basepairs of the fatty acid synthase (FAS) promoter (X54671; basepairs 1300 to 1600). It should be noted that the reduction of matches mainly affects redundant and false positive matches.

Search mode	Number of matches
Threshold = 0.85 / individual matrices	155
Threshold = 0.85 / families	115
Optimized matrix threshold / individual matrices	64
Optimized matrix threshold / families	46

Table 8-9. Number of MatInspector matches with different settings on 300bps of the FAS promoter.

8.7.3 Analyzing promoters for common transcription factor binding sites

Although MatInspector can find most true positive and only few false positive TF binding sites in a promoter region, it is not necessarily clear if these sites are functional in a wet lab experiment. Potential binding sites in a promoter can be functional in certain cells, tissues or development stages only. A first step in elucidating functionality is a comparative promoter analysis, i.e. to analyze promoters that share a common function, for example promoters that are responsive to interferon or a set of actin promoters from different species. If a binding site occurs in most promoters, maybe even at a similar relative position within the promoters, this is supporting evidence that the individual sites might be functional. Often the analysis for TFBS common to an appropriate set of promoters can also lead to other new findings and hypotheses.

As a first step to in-depth promoter analysis this functionality together with a convenient graphical display of the results was added to MatInspector's capabilities lately. In the following example a set of 11 promoter sequences for muscle actin genes was analyzed for all TF binding sites identified by MatInspector that are common to the input sequences.

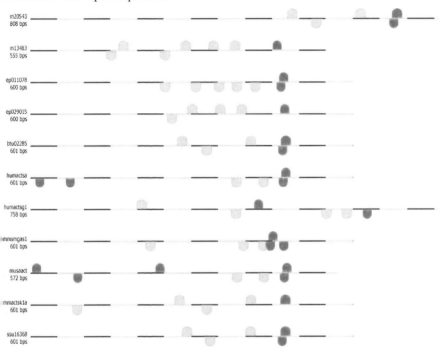

Figure 8-3. *Transcription factor binding sites common all sequences in a set of 11 muscle actin promoters. Different colours denote different transcription factor families (light grey: Serum Response element binding Factor / SRF, dark grey: Tata-Binding Protein Factor / TBPF).*

This simple analysis for common TF sites in coregulated promoters can reduce the number of relevant TFs dramatically. Additionally Figure 8-3 already shows a pattern of TFBSs in muscle actin promoters, which consists of a number of SRF sites and a TATA box which is conserved throughout the sequences. Often the results are not as clear as in this example and it might be necessary to reduce the quorum restraint, i.e. the number of sequences that must contain the TF site, because of input sequences that are regulated by a different mechanism.

8.8 Organizational promoter models

As stated before the regulation of genes is precisely controlled by the different combinations of TF-bound sites in various cell types. A prime example is the regulation of the RANTES/CCL5 gene[12]: The 300 basepair promoter of this chemokine contains six TF binding regions that were assessed experimentally in 5 human cell types under both stimulated and unstimulated conditions. It was shown that various subsets of the six binding regions each containing several TFBSs play a role in the regulation of the gene in different tissues. Each of the TFBSs was relevant for some tissues but nonessential in others as was determined by mutation experiments.

Therefore, identification of individual matrix matches is usually only the first step in promoter analysis, the subsequent aim will be the identification of more complex promoter models, i.e. functional units of a promoter consisting of at least two TFBS in conserved order. So-called promoter modules form a functional unit, allowing synergistic or antagonistic effects for a specific activation or repression of a gene.

NFkappaB, for example, is a transcription factor that is known to regulate various genes such as Interleukin2 (IL2) and Interferon-beta (IFNβ). However, NFkappaB alone is not sufficient for the activation of these genes; it requires at least a second TF, such as CREB for the regulation of Interleukin[13] or the factor IRF1 for the regulation of HLA-A and HLA-B[14]. The correct regulation of interferon-beta in different pathways requires two different modules in its promoter consisting of NFkappaB and CREB[15] on one hand as well as NFkappaB and IRF[16] on the other hand.

The module in Figure 8-4 confers inducibility by tumor necrosis factor alpha (TNF-alpha) and gamma-Interferon (gamma-IFN) to several promoters of the MHC/HLA class I genes as well as to beta-2 microglobulin and beta-Interferon genes and can be detected by comparative promoter analysis based on MatInspector [17].

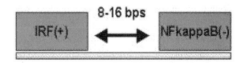

Figure 8-4. Promoter module found in MHC, beta-2 microglobulin and beta-Interferon, showing order, strand orientation and distance between anchors of the single sites as found by MatInspector.

For the automatic detection of modules and more complex models only those combinations of TF sites are analyzed that are situated in a common order and distance from each other. This requires a program that takes the single TFBS found by MatInspector as input and analyses their organization more thoroughly. This program called *FrameWorker* is available at www.genomatix.de within a program package called GEMS Launcher. Since a generic algorithm to find all possible subsets (longest common subsequence problem) is np-complete and thus computationally expensive, some restrictions have to be set as parameters.

Setting one of the restraints, the minimum and maximum distance between the anchors of individual TF sites to 10 and 100 basepairs respectively in the actin example from above yields the model depicted in Figure 8-5. The common pattern automatically found in the muscle actin set consists of two SRF sites in a distance of 33 to 91 basepairs, followed by a TATA box in 13 to 87 basepairs distance. This promoter model is part of a more complete muscle specific actin promoter model consisting of seven elements including SRF, TATA and SP1 as described and evaluated in [18].

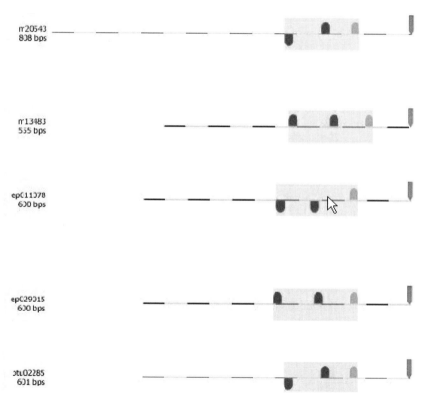

Figure 8-5. Matches to a common promoter module found in at least 90% of the muscle actin promoter set, shown for 5 of the 11 promoters. SRF and TATA sites are depicted in dark grey and light grey respectively, the transcription start site is marked by a red arrow.

179

This promoter model can now be used to scan DNA sequences (e.g. the complete human genome) for the occurrence of similar patterns. When searching a database of 5,9199 human promoters (35 million basepairs), 347 of them contain the model including the promoters for alpha1 actin, alpha2 actin, gamma2 actin, myosin and carbonic anhydrase. In contrast to searching the promoter database for single matches to SRF or TATA resulting in 33,692 and 96,166 hits respectively this shows a higher selectivity by orders of magnitude. For this analysis the program package GEMS Launcher at www.genomatix.de was used.

8.9 Regulatory networks

Like the well-known maps of metabolic networks that describe pathways potentially used by a cell to accomplish metabolic processes, regulatory networks describe regulator-gene interactions that show potential pathways a cell can use to control gene expression. Knowing at least parts of these networks is important for example to understand the molecular underpinnings of cell life, which may later on help to eliminate side effects when developing new drugs.

MatInspector can find the various combinations of activators and repressors that bind to specific DNA regulatory sequences governing the transcription of eukaryotic genes. Especially the concurrent expression of genes as observed in expression array analysis in part is orchestrated by common regulatory elements. These regulatory modules can be discovered as shown before and used to identify additional target genes with similar regulatory properties in genomic sequence databases. It should be stressed that such an analysis is complicated by the fact that coexpressed genes in an expression array experiment are not necessarily all coregulated, as different regulation mechanisms can lead to the same expression pattern. Effects like a secondary cascade of transcription activation during the time course can divide a coexpressed cluster of genes in subsets regarding coregulation. The careful selection of gene clusters is a crucial step for a successful promoter analysis.

Promoter modules provide the basis to elucidate regulatory networks involved in gene expression, because they represent the molecular basis for integration of several TF signals into one output (transcription or repression).

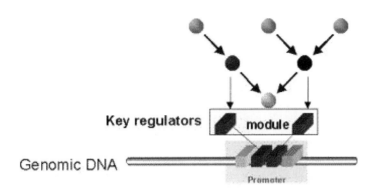

Figure 8-6. *Elucidating regulatory networks by promoter analysis.*

For Saccharomyces cerevisiae more or less complete regulatory networks have been identified and published by now[19,20]. Deducing corresponding networks for higher eukaryotes such as mouse or human might turn out more difficult as there are some obstacles to overcome[21]: First the promoter sequences of higher eukaryotes are much harder to obtain as they might be located kilobases upstream of their corresponding reading frames because of non-coding leader exons and introns. Second, human promoters are usually more complex than yeast promoters and use very different sets of TFs as illustrated in the RANTES/CCL5 example above. But since the general principles of promoter organization are the same in yeast and humans, and promoter sequences for higher eukaryotes are becoming available lately (e.g. [22]) it is a question of time when a complex and complete regulatory network for human genes will be available. Although we have learned in the meantime that finding of TFBSs MatInspector style will not be sufficient to understand gene regulation, it will remain a crucial step in the chain of analytical events finally leading to understanding of regulatory networks on a molecular level.

Acknowledgments

I would like to thank K. Frech and especially T. Werner for critically reading the manuscript and their helpful suggestions.

References

1. Quandt, K., Frech, K., Karas, H., Wingender, E., Werner, T.: MatInd and MatInspector: new fast and versatile tools for detection of consensus matches in nucleotide sequence data. *Nucleic Acids Res.* 1995 Dec 11;23(23):4878-84.
2. Lander, E.S., Linton, L.M., Birren, B., Nusbaum, C., Zody, M.C., Baldwin, J., Devon, K., Dewar, K., Doyle, M., FitzHugh, W., Funke, R., Gage, D., Harris, K., Heaford, A., Howland, J., Kann, L., Lehoczky, J., LeVine, R., McEwan, P., McKernan, K., Meldrim, J., Mesirov, J.P., Miranda, C., Morris, W., Naylor, J., Raymond, C., Rosetti, M., Santos, R., Sheridan, A., Sougnez, C., Stange-Thomann, N., Stojanovic, N., Subramanian, A., Wyman, D., Rogers, J., Sulston, J., Ainscough, R., Beck, S., Bentley, D., Burton, J., Clee, C., Carter, N., Coulson, A., Deadman, R., Deloukas, P., Dunham, A., Dunham, I., Durbin, R., French, L., Grafham, D., Gregory, S., Hubbard, T., Humphray, S., Hunt, A., Jones, M., Lloyd, C., McMurray, A., Matthews, L., Mercer, S., Milne, S., Mullikin, J.C., Mungall, A., Plumb, R., Ross, M., Shownkeen, R., Sims, S., Waterston, R.H., Wilson, R.K., Hillier, L.W., McPherson, J.D., Marra, M.A., Mardis, E.R., Fulton, L.A., Chinwalla, A.T., Pepin, K.H., Gish, W.R., Chissoe, S.L., Wendl, M.C., Delehaunty, K.D., Miner, T.L., Delehaunty, A., Kramer, J.B., Cook, L.L., Fulton, R.S., Johnson, D.L., Minx, P.J., Clifton, S.W., Hawkins, T., Branscomb, E., Predki, P., Richardson, P., Wenning, S., Slezak, T., Doggett, N., Cheng, J.F., Olsen, A., Lucas, S., Elkin, C., Uberbacher, E., Frazier, M., Gibbs, R.A., Muzny, D.M., Scherer, S.E., Bouck, J.B., Sodergren, E.J., Worley, K.C., Rives, C.M., Gorrell, J.H., Metzker, M.L., Naylor, S.L., Kucherlapati, R.S., Nelson, D.L., Weinstock, G.M., Sakaki, Y., Fujiyama, A., Hattori, M., Yada, T., Toyoda, A., Itoh, T., Kawagoe, C., Watanabe, H., Totoki, Y., Taylor, T., Weissenbach, J., Heilig, R., Saurin, W., Artiguenave, F., Brottier, P., Bruls, T., Pelletier, E., Robert, C., Wincker, P., Smith, D.R., Doucette-Stamm, L., Rubenfield, M., Weinstock, K., Lee, H.M., Dubois, J., Rosenthal, A., Platzer, M., Nyakatura, G., Taudien, S., Rump, A., Yang, H., Yu, J., Wang, J., Huang, G., Gu, J., Hood, L., Rowen, L., Madan, A., Qin, S., Davis, R.W., Federspiel, N.A., Abola, A.P., Proctor, M.J., Myers, R.M., Schmutz, J., Dickson, M., Grimwood, J., Cox, D.R., Olson, M.V., Kaul, R., Raymond, C., Shimizu, N., Kawasaki, K., Minoshima, S., Evans, G.A., Athanasiou, M., Schultz, R., Roe, B.A., Chen, F., Pan, H., Ramser, J., Lehrach, H., Reinhardt, R., McCombie, W.R., de la Bastide, M., Dedhia, N., Blocker, H., Hornischer, K., Nordsiek, G., Agarwala, R., Aravind, L., Bailey, J.A., Bateman, A., Batzoglou, S., Birney, E., Bork, P., Brown, D.G., Burge, C.B., Cerutti, L., Chen, H.C., Church, D., Clamp, M., Copley, R.R., Doerks, T., Eddy, S.R., Eichler, E.E., Furey, T.S., Galagan, J., Gilbert, J.G., Harmon, C., Hayashizaki, Y., Haussler, D., Hermjakob, H., Hokamp, K., Jang, W., Johnson, L.S., Jones, T.A., Kasif, S., Kaspryzk, A., Kennedy, S., Kent, W.J., Kitts, P., Koonin, E.V., Korf, I., Kulp, D., Lancet, D., Lowe, T.M., McLysaght, A., Mikkelsen, T., Moran, J.V., Mulder, N., Pollara, V.J., Ponting, C.P., Schuler, G., Schultz, J., Slater, G., Smit, A.F., Stupka, E., Szustakowski, J., Thierry-Mieg, D., Thierry-Mieg, J., Wagner, L., Wallis, J., Wheeler, R., Williams, A., Wolf, Y.I., Wolfe, K.H., Yang, S.P., Yeh,

R.F., Collins, F., Guyer, M.S., Peterson, J., Felsenfeld, A., Wetterstrand, K.A., Patrinos, A., Morgan, M.J., Szustakowki, J., de Jong, P., Catanese, J.J., Osoegawa, K., Shizuya, H., Choi, S., Chen, Y.J.; International Human Genome Sequencing Consortium: Initial sequencing and analysis of the human genome. *Nature.* 2001 Feb 15;409(6822):860-921.

3. Kodadek, T.: Mechanistic parallels between DNA replication, recombination and transcription. *Trends Biochem. Sci.* 1998 Feb;23(2):79-83.

4. Roulet, E., Bucher, P., Schneider, R., Wingender, E., Dusserre, Y., Werner, T., Mermod, N.: Experimental analysis and computer prediction of CTF/NFI transcription factor DNA binding sites. *J. Mol. Biol.* 2000 Apr 7;297(4):833-48.

5. Werner, T.: The Role of Transcription Factor Binding Sites in Promoters and Their In Silico Detection; in Krawetz SA, Womble DD (eds): *Introduction to Bioinformatics A Theoretical And Practical Approach.* Humana Press, Totowa, New Jersey, 2003, pp 523-538.

6. Cavener, D.R.: Comparison of the consensus sequence flanking translational start sites in Drosophila and vertebrates. *Nucleic Acids Res.* 1987 Feb 25;15(4):1353-61.

7. Knuppel, R., Dietze, P., Lehnberg, W., Frech, K., Wingender, E.: TRANSFAC retrieval program: a network model database of eukaryotic transcription regulating sequences and proteins. *J. Comput. Biol.* 1994 Fall;1(3):191-8.

8. Schneider, T.D., Stormo, G.D., Gold, L., Ehrenfeucht, A.: Information content of binding sites on nucleotide sequences. *J. Mol. Biol.* 1986 Apr 5;188(3):415-31.

9. Bennett, M.K., Lopez, J.M., Sanchez, H.B., Osborne, T.F.: Sterol regulation of fatty acid synthase promoter. Coordinate feedback regulation of two major lipid pathways. *J. Biol. Chem.* 199;270(43):25578-83.

10. Neumann, M., Felber, B.K., Kleinschmidt, A., Froese, B., Erfle, V., Pavlakis, G.N., Brack-Werner, R.: Restriction of human immunodeficiency virus type 1 production in a human astrocytoma cell line is associated with a cellular block in Rev function. *J. Virol.* 1995; 69:2159-2167.

11. Soudeyns, H., Geleziunas, R., Shyamala, G., Hiscott, J., Wainberg, M.A.: Identification of a novel glucocorticoid response element within the genome of the human immunodeficiency virus type 1. *Virology* 1993;194:758-768.

12. Fessele, S., Maier, H., Zischek, C., Nelson, P.J., Werner, T.: Regulatory context is a crucial part of gene function. *Trends Genet.* 2002 Feb;18(2):60-3.

13. Butscher, W.G., Powers, C., Olive, M., Vinson, C., Gardner, K.: Coordinate transactivation of the interleukin-2 CD28 response element by c-Rel and ATF-1/CREB2. *J. Biol. Chem.* 1998 Jan 2;273(1):552-60.

14. Johnson, D.R., Pober, J.S.: HLA class I heavy-chain gene promoter elements mediating synergy between tumor necrosis factor and interferons. *Mol. Cell Biol.* 1994 Feb;14(2):1322-32.

15. Du, W., Thanos, D., Maniatis, T.: Mechanisms of transcriptional synergism between distinct virus-inducible enhancer elements. *Cell.* 1993 Sep 0;74(5):887-98.

16. Leblanc, J.F., Cohen, L., Rodrigues, M., Hiscott, J.: Synergism between distinct

enhanson domains in viral induction of the human beta interferon gene. *Mol. Cell Biol.* 1990 Aug;10(8):3987-93.

17. Klingenhoff, A., Frech, K., Quandt, K., Werner, T.: Functional promoter modules can be detected by formal models independent of overall nucleotide sequence similarity. *Bioinformatics.* 1999 Mar;15(3):180-6.

18. Klingenhoff, A., Frech, K., Werner, T.: Regulatory modules shared within gene classes as well as across gene classes can be detected by the same in silico approach. *In Silico Biol.* 2002;2(1):S17-26.

19. Pilpel, Y., Sudarsanam, P., Church, G.M.: Identifying regulatory networks by combinatorial analysis of promoter elements. *Nat Genet.* 2001 Oct;29(2):153-159.

20. Lee, T.I., Rinaldi, N.J., Robert, F., Odom, D.T., Bar-Joseph, Z., Gerber, G.K., Hannett, N.M., Harbison, C.T., Thompson, C.M., Simon, I., Zeitlinger, J., Jennings, E.G., Murray, H.L., Gordon, D.B., Ren, B., Wyrick, J.J., Tagne, J.B., Volkert, T.L., Fraenkel, E., Gifford, D.K., Young, R.A.: Transcriptional regulatory networks in Saccharomyces cerevisiae. *Science.* 2002 Oct 25;298(5594):799-804.

21. Werner, T.: The promoter connection. *Nat. Genet.* 2001 Oct;29(2):105-6.

22. Scherf, M., Klingenhoff, A., Werner, T.: Highly specific localization of promoter regions in large genomic sequences by PromoterInspector: a novel context analysis approach. *J. Mol. Biol.* 2000 Mar 31;297(3):599-606.

Chapter 9

Recognition of Transcription Factor Binding Sites in DNA Sequences

Evgeny Cheremushkin[1,2], Tatiana Konovalova[1,3],
Tagir Valeev[1,2], Alexander Kel[1,4]

[1]BioRainbow Group, Novosibirsk, Russia

[2]Institute of Informatics Systems, Novosibirsk, Russia

[3]Institute of Cytology and Genetics, Novosibirsk, Russia

[4]BIOBASE GmbH, Wolfenbüttel, Germany

Corresponding author: Evgeny Cheremushkin
630090 room 222 building 6
Lavrentiev ave., Novosibirsk, Russia
Email: cher@biorainbow.com
web-page: www.biorainbow.com

9.1. INTRODUCTION

Gene expression of eukaryotic organisms is regulated mainly by means of multiple regulatory proteins - transcription factors (TF), acting through specific regulatory sequences (TF binding sites) that are usually located in gene promoters, or at more remote locations when acting as part of an enhancer. In multicellular eukaryotic organisms, in order to reflect the huge number of different possible intracellular molecular states, gene expression is provided by means of combinatorial regulation. Combinatorial regulation of transcription is organized by multiplicity of transcription factors (several thousand of different TFs) acting in the cell and interacting with their target sites in DNA, with each other and with particular components of the basal transcription complex as well as with coactivators/corepressors, histone acetylases/deacetylases, therefore making up function-specific multiprotein complexes that are often referred to as enhanceosomes.

Having available genomic sequences on one hand and the massive though phenomenological gene expression data on the other hand, the challenge is to understand regulatory mechanisms of every single gene in the genome by computer analysis of the gene regulatory sequences and by integrating this data with other biological knowledge.

To achieve this goal a lot of attention is paid now to development of computational tools devoted to detailed analysis and recognition of TF binding sites (TFBS) in regulatory regions of genes of different organisms[1].

Number of widespread approaches for recognition of TFBS includes methods of construction of consensus sequences[2-5], weight matrixes[2], oligonucleotide matrixes[6], an estimation of physical and chemical properties[7], construction of the informational measures[8,9], neural networks[10] and various statistical models[11], etc.

Despite of the variety of approaches, the problem of construction of exact algorithms for recognition of TFBS cannot be considered as finally solved. The problems come from the huge variety of contextual, physical and chemical and conformational parameters of real TFBS in different genes; mechanisms of DNA-protein interactions between TFBS and transcription factors and other components of the transcription machinery; specificity of the DNA context surrounding TFBS; extent of evolutionary conservations of the regulatory regions under study.

Therefore a number of different algorithms have been created for recognition of TF binding sites taking into account various specific types of molecular-biological and genomic data related to gene regulation and generated by different experimental approaches.

9.2. LIST OF TERMS USED

DNA Sequence (Region, a fragment or a section of DNA, sequence, DNA)	Sequence in 4-letter alphabet {A, C, G, T}. DNA sequences represent parts of genomes and can contain genes encoding and expressing proteins as their basic function.
Regulatory sequence (regulatory region or a section)	A sequence carrying out a regulatory function. Regulatory function is a circumstantially dependent mediation of possessing of basic functions of genes and other parts of genome. For example, the level of gene expression is regulated by promoters and enhancers – sequences in the 5' upstream regions of genes. About 10-80 % of the human genome corresponds to the regulatory regions of various types.
The base pair (a nucleotide, bp)	One element of DNA sequence, one letter from {A, C, G, T}.
Cis-element (transcription factor (TF) binding site, TFBS, a binding site, a site)	A specific DNA subsequence, 5-30 bp long. These are the "hot spots" of the regulatory function of a region. Cis-elements are often serving as specific sites on DNA for binding of transcription factors — regulatory proteins that are responsible for maintaining the regulatory functions.
Recognition of cis-elements (search for cis-elements, search for sites)	Under recognition we mean a computer search for potential cis-elements in DNA using certain algorithm.
Coding regions of DNA (coding sequence, CDS)	Sections of DNA in genes that correspond to the protein fragments, which a gene encodes. In high eukaryotic genomes the coding regions of genes are interrupted by non-coding regions that correspond to introns in the pre-spliced mRNA molecules that are transcribed from these genes.
Polypeptides, Proteins	Sequences in 20-letter alphabet. Proteins carry out structural, regulatory, catalytic, protective, transport and other functions. Some proteins are complexes of several independent polypeptides, each is encoded by a separate gene.
Gene	A specific set of sequences of DNA that participates in manufacture of a certain protein. This set contains coding, non-coding and regulatory sequences.

Regulation of transcription	The first and most important step in gene expression is gene transcription – synthesis of mRNA on the DNA template. Transcription is regulated (accelerating – activation, slowing down – repression up to complete switching off) through binding of transcription factors to their binding sites in promoters and enhancers and maintaining the regulatory interactions with the RNA polymerase complex.
Level of expression (or expression)	A numerical value reflecting an amount of protein (or the mRNA) produced from the given gene under certain conditions. Often, data are represented in the form of relative expression of a gene under two different conditions (e.g. disease versus normal).
Results of microchip/microarray experiment, or (gene expression profile)	Specific data representing expression levels for a big set of genes (from 5000 genes and above). This data are produced using modern microchip/microarray technologies.

9.3. Molecular Features of Transcription Factors

Transcriptional factors are a large family of regulatory proteins[12,13]. They interact with the DNA of promoters and enhancers in a more or less sequence-specific manner, recognizing defined sequence patterns and/or structural features. In contrast to prokaryotes, where the major control mechanism is to repress a generally active transcription machinery, eukaryotes have to match much more complex requirements to coordinate the execution of genetic programs. This is achieved by directed activation of those genes whose products are needed under certain cellular conditions, in general only a few percent of all genes of the genome. Once bound to the DNA, these factors may influence transcription through several mechanisms:

i)   in most cases studied so far, they enhance the formation of the pre-initiation complex at the TATA-box / initiator element through interaction of a trans-activation domain with components of the basal transcription complex, either directly or through co-activators / mediators;

ii) some transcription factors cause alterations in the chromosomal architecture, rendering the chromatin more accessible to the RNA polymerase(s);

iii) some are auxiliary factors, adjusting an optimal DNA conformation for the activity of another transcription factor;

iv) some factors exert repressing influences, either directly by an active inhibiting domain, or by disturbing the required ensemble of transcription factors within a regulatory array (promoter, enhancer);

189

(v) finally, there is a group of transcription factors that do not directly bind to DNA but rather assemble into higher order complexes through protein-protein interactions.

A definition of "transcription factors" has been proposed earlier[14]: A transcription factor is a protein that regulates transcription after nuclear translocation by specific interaction with DNA or by stoichiometric interaction with a protein that can be assembled into a sequence-specific DNA-protein complex.

Most transcription factors are modularly composed. They may comprise

(a) a DNA-binding domain (DBD);
(b) an oligomerization domain, since most factors bind to DNA as dimers, some also as higher order complexes; in most cases, this region forms a functional unit with the DBD;
(c) a trans-activation (or trans-repressing) domain, which is frequently characterized by a significant overrepresentation of a certain type of amino acid residue (e. g., glutamine-rich, proline-rich, serine-/threonine-rich or acidic activation domains);
(d) a modulating region which is often the target of modifying enzymes, mostly protein kinases;
(e) a ligand-binding domain.

The modular structure of transcription factors and a variety of other TF properties allows them, due to DNA-protein and protein-protein interactions, to control all aspects of a gene-specific transcriptional regulation.

9.4. Existing Types of Biological Data

There are several types of molecular-biological and genomic data related to gene regulation and generated by different experimental approaches and stored in many databases[15]. Various bioinformatics algorithms for predicting of cis-regulatory elements in genome use these data. The quality of prediction depends very much on the amount of information taken by the algorithms from the data sources. Let us consider the main types of the data used.

9.4.1 TF binding sites

TF binding sites participating in regulation of transcription of the corresponding genes are studied through a number of molecular-biological experiments such as DNase I footprinting, gel retardation, gel shift competition, supershift (antibody binding), methylation interference, functional assays with plasmid construction as well as such modern methods as Chromatin IP (CHIP assay). Information is stored in such well known databases as TRANSFAC[16], TRRD[17] and some others. At present TRANSFAC database contains information on more than 10000 TF binding sites in different genes of several eukaryotic organisms (mainly human, mouse, rat). This information is used then to construct methods for prediction of novel TF binding sites in genome. Some such methods will be described bellow.

9.4.2 Promoters, enhancers and silencers

Most of known TF binding sites are located in promoters of genes they are regulating. In Figure 9-1 you can see the distribution of sites in promoters as they are annotated in the TRANSFAC database. Most of the sites are located in rather short region from –300 till +100 of the transcription start. On the other hand, there are many known sites which are located in the rather distant 5' upstream or 3' downstream regions of genes (up to several hundreds kb). They are often organized in "site islands" possessing specific regulatory function enabling activation (enchancers) or repression (silencers) of the genes. Information on known promoters is stored in such databases as EPD (http://www.epd.isb-sib.ch), TSSDB (http://dbtss.hgc.jp) and others. Knowledge whether the sequence under study belongs to a promoter or enhancer of a certain gene can be used to predict functional cis-regulatory elements.

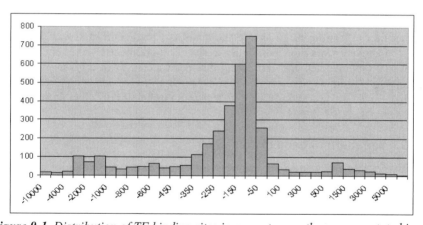

Figure 9-1. Distribution of TF binding sites in promoters as they are annotated in TRANSFAC database TRANSFAC 8.1 (2004-03-31).

9.4.3. Homology of regulatory sequences between different organisms

Comparative genomics provide a rich source of information. It allows to reveal information on evolutionary homology between regions of genomic sequences of different organisms (so called orthologous regions). Such regions often contain orthologous genes including their protein coding regions as well as regions covering introns, promoters, or far 5' and 3' regions of genes. Often, these regions contain regulatory genomic sequences possessing evolutionary homology between different organisms. For instance, by comparison of human to mouse genome 2262 highly conserved noncoding regions were revealed in these genomes[18]. It is tempting to speculate that these regions contain functionally important regulatory sequences, which remain conserved in evolution. Using this kind of comparative genomic data we improve the precision of TF binding sites prediction in genomes.

9.4.4. Functionally related regulatory sequences

Functional annotation of genomes gives rise to creation of functional ontologies that group genes and their products into functionally related categories. Gene ontology (GO) is most widely used ontology of such type. It helps to group genes according to the function of their products or their location in the cell, or according to their structural properties. For example, genes involved in one signal transduction pathway, metabolic circuit or other biological process can be grouped together. A coordinated regulation of expression of the functionally related genes is provided by common transcription factors and specific combinations of their binding sites in promoters of these genes. Such combinations of sites situated in a close proximity to each other are called composite elements (CE) or composite regulatory modules[19, 20] and serve as specific targets for synergistic binding of several transcription factors forming so called enhanceosomes[21] through numerous protein/DNA and protein/protein interactions. Account of such composite modules in promoters and enhancers can significantly increase the quality of prediction of single TF binding sites. Moreover, it provides possibility to specify more precisely the functional role of the TF binding sites via considering the specific context of a given promoter. A number of algorithms were developed so far for analysis of such composite modules. Hereafter we will describe software tools that use principles of "Genetic algorithms" in order to accomplish this task.

9.4.5. Pattern of gene expression

Using latest advances in microarray (microchip) technologies molecular biologists get very powerful means to study patterns of expression of huge number of genes (up to the complete set of genes from one genome) in parallel manner in many different conditions (different tissues, cell types; consecutive time points in cell response on to an external signal or drug; disease state versus normal tissues and so on)[22]. Accuracy of these methods at present is not so high, but sufficient for wide and an effective use of this information. Numerous statistical approaches have been suggested for analysis of microarray gene expression data[23]. Most of them lead to revealing of "gene clusters" — sets of genes that can be characterized by a common pattern of expression changes in various conditions. It is generally accepted that genes in such clusters should be regulated through some common molecular mechanism. So, the same bioinformatic approaches as for the functionally related genes can be used here. The characteristic feature of such kind of data is a huge scale (in the number of genes) and high level of experimental noise.

9.4.6. Single-nucleotide polymorphism

Single-nucleotide polymorphism (SNP) — is a difference in the DNA sequence between different individuals of one species. Latest studies show that in human populations in average one polymorphic DNA position appears in each 1000 nucleotides of genome sequence[24]. SNPs are responsible for individual differences of individuals in morphology, physiology and behavior. Polymorphisms in regulatory regions of genes can influence their regulation. Thus, using corresponding algorithms, there is an opportunity to predict change of regulation in relation to SNPs.

9.4.7. Genomic sequences with unknown function

To less than 20% of the human genome we can assign a certain function. According to all available information[25], about 50% of genomic DNA is transcriptionally active. This means that a lot more regulatory sequences are present in the genome than it is known at present. It is a great challenge to analyze genomes with appropriate bioinformatic tools enabling us to integrate all available information and to make prediction of potential regulatory sequences in the genome.

9.5. Algorithms of Cis-elements Recognition

In this section we shall describe several algorithms for prediction and analysis of TF binding sites. Among them there are: Match algorithm for searching sites by means of weight matrixes; algorithm for searching double sites; phylogenetic footprint — an algorithm for searching of evolutionally conserved sites; a context-dependent method for site search.

9.5.1. Weight matrix algorithm

The main concept of a weight matrix algorithm[2, 26] is to assign four weight values for every site position corresponding to the four nucleotides A, T, G and C. These weights represent frequencies of each nucleotide occurrence in the current position. Let $F = |f_{ij}|$ be a 4xl-matrix of nucleotide frequencies, f_{ij} is absolute frequency of the i-th nucleotide occurrence in the position j of learning set, which consists of aligned nucleotide fragments comprising well-known binding sites for a considered transcription factor (i=1,...,4; j=1,...,L). Now we can define the weight matrix W. There are many approaches to define its elements λ_{ij}. The following approach is one of the most frequently used (Table 9-1):

$$\lambda_{ij} = I(i) * f_{ij} \quad , \text{ where } \quad I(i) = \left| \sum_{B \in \{A,C,G,T\}} f_{i,B} * \ln(4 * f_{i,B}) \right| \quad ,$$

193

Site position		1	2	3	4	5	6	7	8
F:	'A'	13	14	2	23	24	1	0	1
	'C'	4	3	7	0	0	23	1	1
	'G'	4	5	1	0	0	0	17	21
	'T'	3	2	14	1	0	0	6	1
W:	'A'	1053	1161.3	170.5	2423.9	2629.1	105.4	0	97.2
	'C'	324	248.8	596.6	0	0	2423.9	92.2	97.2
	'G'	324	414.7	85.2	0	0	0	1567.2	2041.3
	'T'	243	165.9	1193.2	105.4	0	0	553.1	97.2

Table 9-1. *Example of frequency (F) and weight (W) matrix.*

The method for recognition of new potential TF binding site in any given fragment of DNA of L nucleotides in length using weight matrix W is as outlined below.

For the fragment a score value w is calculated as follows:

$$w = \frac{x - x_{min}}{x_{max} - x_{min}} \ ,$$

where

$$x_{max} = \sum_{j=1}^{L} \max_i (\lambda_{ij}), \quad x_{min} = \sum_{j=1}^{L} \min_i (\lambda_{ij}) \ ,$$

while x represents similarity between given sequence and learning set:

$$x = \sum_{i=1}^{L} \lambda_{i,B} \ , \quad B \in \{A, C, G, T\}$$

If w is higher then a predefined threshold $w_{cut-off}$ the potential TF binding site is predicted by the program. All potential sites are predicted in a given nucleotide sequence by applying the described algorithm to every position of sliding window on the sequence.

Exercise 1

1. Download and install software package 'Regulatory Sequence Analysis Tool' from http://www.biorainbow.com/reg_seq_analysis/ .
2. Select algorithm 'Match', push the button 'Select Matrix File', select demo_matrixlib.dat. Then press 'Select Sequence File' and select demo_promoters.embl. In the sequence list select sequence SELE. This sequence corresponds to the promoter of human gene SELE.
3. Push 'Calculate'. The results window should appear. Decrease cut-off to 0.70. You will see two sites found in the sequence with given cut-off value. Click on the first, then on the second and you will see additional information about these sites. (See Figure 9-2).

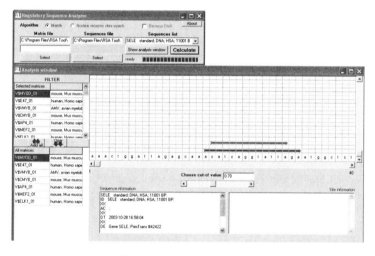

Figure 9-2. *Example for the Exercise 1.*

9.5.2. Double sites recognition algorithm

Binding sites of some transcription factors (e.g. nuclear receptors) consist of two so called "half-sites" or "repeats" with varying distance between them. The distance depends on the type of the factor that recognizes this site. Half-sites can have similar structure. Since the site consist of two conservative domains with varying distance (Figure 9-3.), let's define double-site recognition models M_k represented in the Figure 9-3.

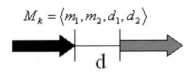

Figure 9-3. *Site consists of two conservative domains with varying distance between them.*

Here, m_1 and m_2 are weight matrices[26], while d_1 and d_2 are minimal and maximal distance between half-sites. Assume that w_1 *(i)* and w_2 *(j)* are the scores of m_1 and m_2 in sequence positions *i* and *j* respectively. A site is considered as recognized when an average score value $w = \dfrac{w_1(i) + w_2(j)}{2}$ exceeds given cut-off ?, and distance between half-sites $d \in [d_1, d_2]$.

Due to the unknown varying distance between half sites it becomes difficult to build an appropriate model based on a set of known sites of this type. We suggest the following algorithm of building this model, which is based on an iterative procedure of applying Gibbs Sampling method[27] and subclustering the set of sites.

Let $S = (S_1,...,S_m)$ be a training set of sequences of TF binding sites. A basic algorithm of model construction uses a subset of sequences from the training set as follows. For each subset $S' = (S_1',...,S_m')$ of the set S we can choose two sets of short subsequences $S^1 = (s_1^1,...,s_n^1)$ and $S^2 = (s_1^2,...,s_n^2)$ $s_i^1, s_i^2 \in S_i'$, the length of s_i^j is equal to 6 nucleotides.

Let's find by means of classical Gibbs Sampling procedure[27] S^1 and S^2 such, that sequences s_i^1 are similar to each other, and s_i^2 are similar to each other, but they are maximally different from each other. Using sequences from S^1 and S^2 we create corresponding matrixes m_1 and m_2. Then we choose the distances $d_1 = \min_i \left(d(s_i^1, s_i^2) \right)$ and $d_2 = \max_i \left(d(s_i^1, s_i^2) \right)$.

To initiate the algorithm we choose an initial subset $S_{[0]}$ named a base set. Usually the base set consists of small number of sites with already known structure. Now we build a model $M_{[0]}$ using the base set only. Then, we apply the model $M_{[0]}$ on the sequences that were not included into the base set $S_{[0]}$. One sequence with maximal score $w_{[0]}$ of the model $M_{[0]}$ is added then into the initial set to form the set $S_{[1]}$. Based on the new set the new model $M_{[1]}$ is constructed using the basic algorithm described above. We continue the procedure of addition of new sequences until the score $w_{[k]}$ exceeds the threshold *c*. (Figure 9-4).

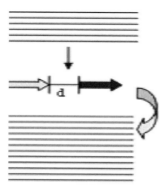

Figure 9-4. *Let's choose an initial subset* $S_{[0]}$ *- a base set. Build a model* $M_{[0]}$ *and add in* $S_{[0]}$ *a sequence from* $S \backslash S_{[0]}$ *for which the weight* $w_{[0]}$ *of the model* $M_{[0]}$ *is maximal. Thus, we will get a new model* $M_{[1]}$. *Continue procedure of adding new sequences until the weight* $w_{[k]}$ *exceeds initially set cutoff C.*

After the termination of the algorithm we receive the model M describing double sites in the set S.

Exercise 2

1. Download and install software package 'Regulatory Sequence Analysis Tool' from http://www.biorainbow.com/reg_seq_analysis/ , if not already done in Exercise 1.
2. Select algorithm 'Nuclear Receptors Site Search', push the button 'Select Matrix File', select test.clib and then test.matrixlib. Then press 'Select Sequence File' and select demo_promoters.embl. In the sequences list select sequence SELE. This sequence corresponds to the promoter of human gene SELE.
3. Push 'Calculate'. The results window should appear. Decrease cut-off to 0.70 You will see two sites found in the sequence with given cut-off value. Click on the first site, then on the second and you will see additional information about these sites. (See Figure 9-5).

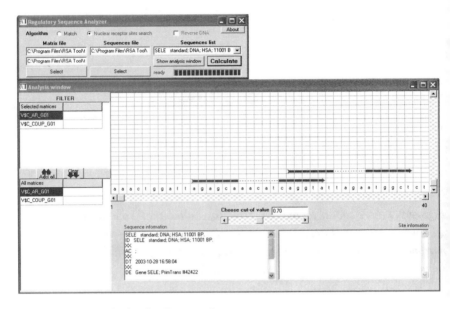

Figure 9-5. *Example for the Exercise 2.*

9.5.3. Phylogenetic footprint

The algorithm of phylogenetic footprint is an algorithm that considers evolutionary processes in regulatory genomic sequences. The idea of the method is based on the assumption, that TF binding sites in promoters should evolve essentially more slowly, than other regions that do not possess any function conserved in evolution. Thus, potential TF binding sites found in evolutionary-conserved regions of promoters could be considered as "real" sites more evidently than other sites found in other genomic regions.

We have developed a novel algorithm for aligning two or more orthologous regulatory sequences. The algorithm is based on the assumption, that during evolution cis-elements are more conserved, than other sections of promoter sequences. The new alignment algorithm is called motif-based alignment[28] and includes information about TF binding sites. To search for the sites we apply position weight matrices (PWM) from the TRANSFAC database (www.biobase.de). Every nucleotide in a sequence can potentially belong to one or several TF binding sites. We estimate the probability $w_p(\bar{S},k)$ of k-th nucleotide of a sequence $\bar{S}$ to belong to a binding site of a factor $T_n(p \in [1,P])$:

$$w_p(\bar{S},k) = \alpha \times \sum_{j=k-L+1}^{k} \exp\left(\beta \times s_p(\bar{S},j)\right), \vec{w}(\bar{S},k) = \left\langle w_1(\bar{S},k),...,w_P(\bar{S},k)\right\rangle$$

where $s_p(\bar{S},j)$ is the score of p-th matrix at j-th position of sequence, L is the length of $\bar{S}$, and α and β are two normalization constants. We use a smoothing function that weights differently the positions of the sites giving more weight to the core positions.

It is known that the library of weight matrices contains matrices that are similar to each other. These are different matrices for the same transcription factor or for transcription factors that are very similar in their DNA binding signature. We consider a similarity matrix M that takes into account similarities between weight matrices. We use M to convert the probability to a new function:
$\vec{\varphi}(\bar{S},k) = \vec{w}(\bar{S},k)\cdot M$, where is P x Q similarity matrix. We will use $\varphi(a)$ instead of $\varphi(\bar{S},k)$, where $a \in \Sigma \times \Phi$ is a sequence element and $\gamma(a) \in \Sigma$ is a nucleotide for this element. The components of the vector $\varphi(a)$ we will call TF belonging coefficients.

The alignment algorithm is similar to the generally accepted Needleman-Wunsch dynamic programming algorithm[29]. A major modification is made in the way of calculating the nucleotide substitution weights and gap penalty. The PWM scores were considered at every sequence positions in order to compute the corresponding substitution weights and gap penalty (see Figure 9-6).

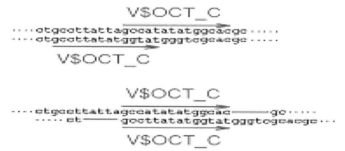

Figure 9-6. We consider alignment as a favorable one, if sites are aligned to each other.

For calculation of the gap penalty we construct a disjoining score function that can be applied to any two neighbor positions a and b in one nucleotide chain:

$$X_{gap}(a,b) = \left(\varphi(a) + \varphi(b)\right)^2$$

.

This score estimates how similar are the two TF belonging vectors for these two positions. If both positions have high belonging coefficients for the similar sets of transcription factors then the score X_{gap} is high. It means that most of the predicted TF binding sites in this region span over these two neighbor positions and disjoining of these positions by a gap will cause braking of these TF binding sites and is considered as an unfavorable event.

199

For N sequences in the alignment we use the following variant of the disjoining score function, where C_{gap}, W_{gap} are optimized constants:

$$Y(a,b) = C_{gap}/N + W_{gap} \cdot X_{gap}(a,b)$$

We use this function for calculation of the gap penalty, while inserting gap in $\bar{S}^1$ between $k-1$ and k under position l in $\bar{S}^2$:

$$GAP(\bar{S}^1, \bar{S}^2, k, l) = \frac{G(\bar{S}^1, k) + R(\bar{S}^2, l)}{2},$$

where

$$G(\bar{S}^1, k) = Y(s^1_{k-1}, s^1_k),$$

which describes the score of disjoining of positions $k-1$ and k due to a possible insertion in the sequence $\bar{S}^2$ during evolution;

$$R(\bar{S}^2, l) = \frac{Y(s^2_{l-1}, s^2_l) + Y(s^2_l, s^2_{l+1})}{2},$$

which describes the score of simultaneous disjoining of positions $l-1$ and l as well as l and $l+1$ due to a possible deletion in the sequence $\bar{S}^1$ during evolution. Both deletions and insertions are considered to be equally probable that is why gap penalty is a mean value between G and R.

Substitution weight for two aligned positions k and l in the sequences $\bar{S}^1$ and $\bar{S}^2$ accordingly:

$$SUB(\bar{S}^1, \bar{S}^2, k, l) = Z(s^1_k, s^2_l),$$

$$Z(a,b) = \frac{\Delta}{N} \cdot C_{sub} - W_{sub} \cdot \sum_{i=1}^{3} \lambda_i \cdot E_i(a,b) \Big/ \sum_{i=1}^{3} \lambda_i, \text{ where, } \Delta = \begin{cases} 1, \gamma(a) \neq \gamma(b) \\ 0, \gamma(a) = \gamma(b) \end{cases},$$

$$E_1(a,b) = \begin{cases} (\bar{\varphi}(a) + \bar{\varphi}(b))^2, \gamma(a) = \gamma(b) \\ \bar{\varphi}(a)^2 + \bar{\varphi}(b)^2, \gamma(a) \neq \gamma(b) \end{cases}, \quad E_2(a,b) = \max_i(\varphi_i(a) \cdot \varphi_i(b)),$$

$$E_2(a,b) = \begin{cases} 0, m > C_{min} \\ (C_{min} - m)/C_{min}, m \leq C_{min} \end{cases}, \text{ where } m = \min_i |\varphi_i(a) - \varphi_i(b)|.$$

$\gamma(a) \in \Sigma$ is nucleotide, C_{sub}, C_{gap}, W_{sub}, W_{gap}, λ_i are constants.

200

In Figure 9-7, we present an example of alignment of two sequences that is done by the motif-based algorithm. The score values of the aligned sequences are shown above and under the sequences accordingly. It can be seen that the peaks of the TF belonging coefficients are aligned to each other.

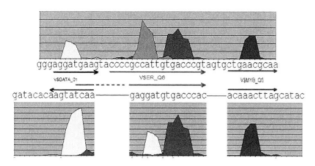

Figure 9-7. *Results of motif-based alignment of two regulatory sequences. TF belonging coefficients are aligned to each other along the sequences. Arrows show potential TF binding sites in both sequences.*

Exercise 3

1. Go to the web-site http://www.biorainbow.com/footprint/
2. Enter name of sequence 'test-sequence'
3. Check 'take an example' checkbox.
4. Press 'submit' button (See Figure 9-8.)

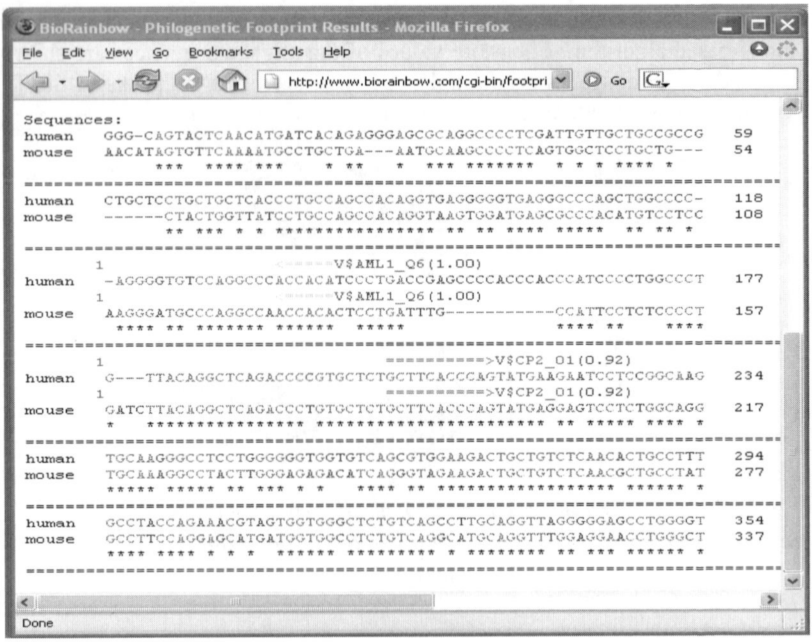

Figure 9-8. *Example for the exercise 3.*

9.5.4. An algorithm for analysis of composite regulatory modules

We introduce an algorithm for revealing composite regulatory modules in a set of regulatory genomic sequences selected by a similarity of the function they possess (for example, promoters of genes preferentially expressed in one tissue, or in one and the same stage of development). Composite modules are characterized by a defined set of transcription factors and a specific location of their binding sites in such promoters and enhancers. It is known, that transcription factors in such composite modules can cooperate with each other and form specific regulatory complexes on DNA. Sites of these factors are usually located in a close proximity to each other.

Let's define the mathematical model for the composite modules (CM) as a set of TF matrices and a set of parameters that define the window w, cut-offs of the matrices c, maximal number of most scoring matches found by these matrices inside the window and a function for computing the composite score. We use a linear function for calculation of the composite score in a sequence S: $F(S) = \sum_{i \in CM} Nsite_i(S)$, where $Nsite_i(S) < Nsite_i$ is the number of sites found in the sequence by the i-th matrix of the CM. $Nmax_i$ is the maximal number of considered most scoring matches found by these matrices.

The algorithm takes two sets of sequences as an input (sequences under analysis and a set of background sequences). The algorithm chooses a CM model that maximizes the discrimination value R = (F+ - F-) / (d + + d-), where F + and F- are mean values, and d + and d- are variances of the composite score values F for the set of sequences under analysis (+) and for the background sets (-) accordingly[20]. Thus, we can find a set of transcription factors controlling given set of genes.

The space of combination of all possible parameters, including the optimal set of matrices, their cut-off values, and the length of the windows, is enormously big. It appeared impractical to search for an optimal solution using the full search in this space. For optimization two approaches are used: Genetic algorithm and algorithm of metropolis.

Genetic algorithm. At the first stage, an initial pool of "genomes" G is created. Each genome G = {S, C, N} describes a composite module and contains a set S of m matrixes with according cut-offs C and maximal number of considered sites N. The initial pool of genomes is generated randomly by random selection of matrices, their cut-offs and other parameters. Thereafter the pool of genomes is considered subject of mutations, as a naturally evolving population, recombination and selection (individual matrices and assigned to then other parameters are considered as "genes" in such evolving population of "genomes"). On each following step of "evolution" some part of genes undergoes mutations by change of matrix or increase/decrease the cut-off and/or maximal number of considered sites. The recombination is carried out by an exchange of "genes" between various genomes. Genomes that have got the maximum discrimination value R are predominantly selected for inclusion into next "generation". After several cycles of such mutations, recombinations and selection the best genome G is defined. It describes a combination of weight matrixes with optimum cut-offs and other parameters specific to the given set of regulatory sequences.

Algorithm of metropolis.
1. Initially at step 0, a composite module G(0) is generated by a random choice of the matrices S(0), their cut-off values C(0) and the maximal number of considered sites N(0). The discrimination value R(0) is calculated.
2. At each consequent step i a new composite module G(i) is created from the composite module G(i-1) by a random "mutation" of one of the parameter (a change of matrix, or change in a matrix cut-off or in maximal number of considered sites). The value R(i) is calculated.
3. The following condition is checked:
   a. If R(i)> =R(i-1), we "accept" the new composite module G(i);
   b. If R (i) <R (i-1), the new module G(i) is accepted with probability R(i)/R(i-1)
4. If the new G(i) is not accepted, the module is copied from the previous step G(i) =G(i-1).
5. Steps 2-4 are repeating many times.

In is known that the sequence of composite modules G(0) … G(n) for the large n characterizes distribution of the function R. A composite module that appears in such sequence most often corresponds to the maximum of the function R. Moreover, more parameters of the distribution can be obtained from this sequence, for example, deviation of each component in the composite module, in order to find out the most stable and more optional parts of the module.

9.5.5. Algorithm of recognition of TF sites using local context

Let $Q=\{q_1,...,q_m\}$ — be a set of "positive" sequences or sequences containing regulatory elements (i.e. promoters) and $T=\{t_1,...,t_k\}$ — be a set of "background" sequences defining the background genomic sequence distribution. Lets also define a block $B=\{b_1,...,b_l\}$ — as a pattern of length l consisting of nucleotides. On a sequence fragment s we define a rule $f(s) \in R$, in such a way that if $f(s) > 0$ then s is recognized as a TF binding site, otherwise, it is not recognized as a TF binding site. Define $f(s)$ in the following way: $f(s) = \sum_{i=0}^{N} f_i(s)$, where $f_i(s) = c_i^1$, if in the region $\left[p_i^1, p_i^2 \right]$ a block B_i exists, and $f_i(s) = c_i^2$, if block B_i does not exist. Let's call $< B_i, p_i^1, p_i^2 >$ a block model.

Using the maximum likelihood criteria we find that $c_i^1 = \log(fr_i^1) - \log(fr_i^2)$, $c_i^2 = \log(1 - fr_i^1) - \log(1 - fr_i^2)$, where fr_i^1 - is a frequency of the block B_i in the $\left[p_i^1, p_i^2 \right]$ region in the sequence set $Q=\{q_1,...,q_m\}$ and fr_i^2 is the frequency of the block B_i in the region $\left[p_i^1, p_i^2 \right]$ in the sequence set .

The algorithm selects a small number N (up to 10) of the best blocks to construct the model $f_i(s)$. The N best blocks are represented in a graphical user interface of the program. The quality of the block model $f_i(s)$ depends on the coefficients c_i^1 and c_i^2. Region $\left[p_i^1, p_i^2 \right]$ is selected from region [-P,P] around the core of the binding site. The length of the flanks P is defined in the graphical interface.

Now, having two sets of sequences on the learning stage we compute the block-model. This model is used then for scanning other sequences to search for the new potential TF binding sites. Such consideration of the local sites context helps to improve the recognition quality as it was shown for slice sites[30] and for TF binding sites[31,32].

Exercise 4

1. Download and install software package 'Block Site Searcher' from http://www.biorainbow.com/bss/.
2. Push the button 'Profile', select demo_profile.prf. Then press 'Matrices Library' and select demo_matrixlib.lib.
3. Push 'Learn' button.
4. Push 'Sequence file' and select test_sequence.embl.
5. Push 'Process'. (See Figure 9-9.)

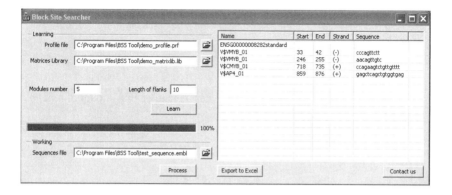

Figure 9-9. *Example for the exercise 4.*

9.6 COMPARISON OF SEARCH ALGORITHMS

The recognition quality can be analyzed from distribution of two parameters: positive prediction value (PPV): $\alpha = 1 - FP$ and sensitivity $\beta = 1 - FN$. They correspond to false positive (FP) and false negative (FN) errors. Let $S = \{s_1,...,s_n\}$ be known experimental sites. $Q = \{q_1,...,q_M\}$ are sites found by a certain method. Denote $s_i \approx q_j$ if site s_i matches q_j (it is recognized by q_j). Let $Q' = \{q_j \in Q \mid \exists s_i \in S, s_i \approx q_j\}$ be a set of sites that recognizes sites from S correctly. $S' = \{s_i \in S \mid \exists q_j \in Q, s_i \approx q_j\}$ is a set of sites that are correctly recognized.

Then $\alpha = \dfrac{|Q'|}{|Q|}, \beta = \dfrac{|S'|}{|S|}$. The basic problem is that not all of the sites are discovered and annotated. Denote $T = \{t_1,...,t_k\}$ as unknown sites. Union of known and unknown sites is the complete set of sites $S^* = S \cup T$. Let's transfer PPV and sensitivity with consideration of unknown sites $\alpha^* = \dfrac{|Q'^*|}{|Q|}$, $\beta^* = \dfrac{|S'^*|}{|S^*|}$ where

$$Q'^* = \{q_j \in Q \mid \exists s_i \in S^*, s_i \approx q_j\}, \; S'^* = \{s_i \in S^* \mid \exists q_j \in Q, s_i \approx q_j\} \cdot$$

We can also rewrite $S'^* = S' \cup T'$ and $Q'^* = Q' \cup Q'_T$. Let unknown sites be k_T times more than known sites: $|T| = k_T |S|$. Let the method recognize smaller percentage of unknown than of known sites:

$$\frac{|Q'_T|}{|Q'|} = k_\alpha \frac{|T|}{|S|} = k_\alpha \cdot k_T \ , \ k_\alpha \in (0,1] \ .$$

Let also the amount of recognized unknown sites depend on the amount of recognized known sites accordingly: $\frac{|T'|}{|S'|} = k_\beta \frac{|T|}{|S|} = k_\beta \cdot k_T \quad k_\beta \in (0,1]$.

We obtain $\alpha^* = \alpha \cdot (1 + k_\alpha \cdot k_T), \ \beta^* = \beta \cdot \dfrac{1 + k_\beta \cdot k_T}{1 + k_T}$.

It is worth to note generally that for various transcription factors there are various constants k_T, k_α, k_β. The constant k_T does not depend on the considered recognition method. As sites from the set T are not known, assume that other constants k_α and k_β also do not depend on the method. Then for the comparative analysis of methods it is enough to use distribution $<\alpha,\beta>$, keeping in mind that it is not an absolute, but relative estimation of methods. The quality of a method for TF site recognition varies for different factors, for different groups of sequences, as well as for parameters of the method. Parameters are not comparable for various groups of factors and groups of sequences, but they are comparable inside one group of factors and sequences.

The system of comparison is implemented in the GRESA DT system as an independent module. It is easy to add a new TF recognition method into this testing module. For this purpose it is enough to implement a new function that uses the common mechanisms of calculation of all the statistics. If the method demands use of some additional data, the data should then be added in such a way that for calculation of statistics the same sets of genes and transcription factors are used. A friendly graphical user interface is provided by this module. (See Figure 9-10).

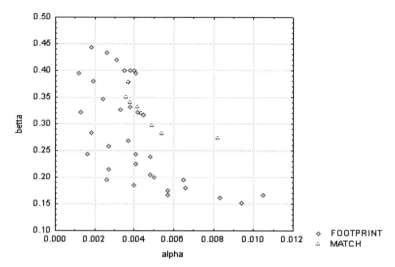

Figure 9-10. *Distribution of positive prediction value* $\alpha = 1 - FP$ *and sensitivity* $\beta = 1 - FN$ *for MATCH and FOOTPRINT methods of binding sites search. The higher values of* α *and* β, *the better quality of recognition. Various combinations of parameters are applied for various practical tasks. Therefore the complete cumulative distribution is shown here.*

9.7 OBJECT-ORIENTED SYSTEM GRESA FOR SEARCH OF CIS-ELEMENTS

GRESA DT environment has hierarchical structure. (Figure 9-11).

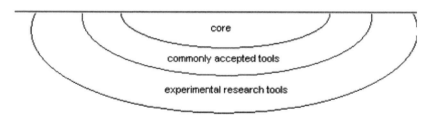

Figure 9-11. *Hierarchical structure of GRESA package. Package consists of three tool levels, core tools, commonly accepted tools and experimental tools.*

The package "core" consists of the classes representing the basic standard objects in bioinformatic science of regulatory DNA sequences:

Sequence – DNA sequence. Represents linear nucleotide sequence, designated by letters A, C, G, T. It also has a name, a description and location in genome (chromosome number, start position on a chromosome and a direction "+" or "-").

Site – a subsequence of DNA sequence, usualy 10-20 b.p., having position, length, and direction.

Factor – an object realizing properties of the transcription factor. Transcription factor is a protein, which binds to a DNA site.

Alignment – a set of the aligned sequences. In each of them gaps between nucleotides can be inserted. Alignments reflect evolutionary similarity of sequences.

Set of sequences, sites, factors – classes in which saving and loading from standard formats is accomplished.

There is also a set of auxiliary classes. Classical operations, such as generation of complementary sequences, sequence searches etc. are also possible.

The set of the standard tools consists of such applications as MATCH[2], COMATCH (search of composite modules), FOOTPRINT[3, 4], and CM SEARCH.

MATCH – a method of site search based on weight matrices. The most widely used method.

COMATCH – a method, searching composite elements and sites with two domains.

FOOTPRINT – a method, which is taking into account evolutionary similarity of sequences.

CM SEARCH – a method of search of the composite modules regulating group of genes. For the given set of genes a common composite module, regulating these genes is searched.

Development and application

The functionality of the GRESA DT environment is growing constantly. Extreme Programming technology for development of the environment makes it possible to support the stable working version. Stability is an important issue having in mind a rather big and distributed group of developers. It is maintained by a large number of team designed automated tests. The life cycle of development of an application consists of several stages. Initially the application is in the stage of experimental development, and then it passes into the stable stage. (Figure 9-12).

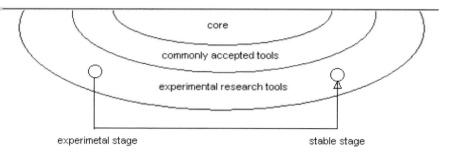

Figure 9-12. Life cycle of the separate application consist of stages when the application is in a stage of experimental development, then passes in a stable stage.

On the later stages it can proceed into a set of standard tools. Any team member can make changes to any class in the source code of the system. The main and only requirement is to keep successful performance of all the tests.

At present GRESA DT is used for regulatory DNA sequence analysis. Application covers the wide range of site recognition tasks. In addition, there are tools for analysis of long regulatory regions such as promoters and enhancers consisting of several TF binding sites. There is an opportunity to construct specific combinations of different methods, for example, phylogenetic footprinting can be done on the basis of TF binding sites predicted by the weight matrix method, or by any other method, or even, it can be done on the basis of other types of regulatory elements (e.g. composite elements). The only requirement is the standard input format. In GRESA DT we have implemented several instruments for comparative testing of different TF recognition methods. This testing system estimates the quality of site recognition.

Exercise 5

1. Download GRESA from http://www.biorainbow.com/gresa/
2. Unpack and open a solution file in Microsoft Visual Studio (version 6.0 or 2003).
3. Open file GresaTest.cpp
4. Study usage of common classes: Sequence, SequenceSet, Site, SiteSet by analyzing the code of test functions.

9.8 CONCLUSION

Sophisticated computational regulatory sequence analysis tools that employ powerful statistical and machine learning algorithms driven by the rich databases that collect biological information enable us to make profound in silico predictions and formulate experimentally testable hypotheses. Such in silico driven experiments can greatly speed up the process of our understanding of gene regulatory mechanisms and the identification of new target genes. The understanding of how gene regulation mechanisms are working with the genomic regulatory sequences will give us powerful means for deciphering causes of major human diseases.

ACKNOWLEDGMENT

The authors are indebted to Olga Kel-Margoulis and Edgar Wingender (BIOBASE GmbH), Aida G. Romashchenko and Vadim A. Ratner (Institute of Cytology and Genetics, Novosibirsk) for fruitful discussion of the results presented in the manuscript. Parts of this work were supported by the Siberian Branch of Russian Academy of Sciences, by a grant of the European Commission (INTAS 03-55- 5218), by Volkswagen-Stiftung (I/75941) grant and by grant BioProfile Braunschweig/Göttingen/Hannover (0313092).

REFERENCES

1. Doolittle RF. Microbial genomes opened up. Nature, 1997; 392: 339–342.
2. Schneider T, Stephens R. Sequence logos: a new way to display consensus sequences. *Nucleic Acids Res.*,1990; 18: 6097-6100.
3. Ulyanov A, Stormo G. Multi-alphabet consensus algorithm for identification of low specificity protein-DNA interactions. *Nucleic Acids Res.*, 1995; 23: 1434-1440.
4. Kel AE, Kondrakhin YV, Kolpakov PhA, Kel OV, Romashenko AG, Wingender E, Milanesi L, Kolchanov NA. Computer tool FUNSITE for analysis of eukaryotic regulatory genomic sequences. Proc. Third Internat. Conf. Intelligent Systems Molec. Biol., 1995: 197-205.
5. Kondrakhin YV, Kel AE, Kolchanov NA, Romashenko AG, Milanesi L. CABIOS., 1995, 9: 1-13.
6. Ponomarenko MP, Ponomarenko JV, Frolov AS, Podkolodnaya OA, Vorobiev DG, Kolchanov NA, Overton C. *Bioinformatics*, 1999; 15: 631-643.
7. Ponomarenko J., Ponomarenko MP, Frolov AS, Vorobiev DG, Overton C, Kolchanov NA *Bioinformatics*, 1999; 15: 654-668.
8. Schneider T, Stormo GD, Gold L. Information content of binding sites on nucleotide sequences. *J. Mol. Biol.*, 1986; 188: 415-431.
9. Papp P, Chattoraj D. Information analysis of sequences that bind the replication initiator Rep A. *J Mol. Biol.*, 1993; 233: 219-230.
10. Horton P, Kanehisa M. An assessment of neural network and statistical approaches for prediction of E. coli promoter sites. *Nucleic Acids Res.*, 1992; 20: 4331-4338.
11. Sewell R, Durbin R. Method for calculation of probability of matching a bounded regular expression in a random data string. *J. Comput. Biol.*, 1995; 2: 25-31.
12. McKnight SL, Yamamoto KR. Transcriptional Regulation. Cold Spring Harbor Laboratory Press, Cold Spring Harbor, 1992.
13. Wingender E. Gene Regulation in Eukaryotes. VCH, Weinheim, 1993.
14. Wingender,E. Classification scheme of eukaryotic transcription factors. *Mol. Biol.* (Mosk) 1997; 31: 483-497.
15. Kel A, Kel-Margoulis O, Borlak J, Tchekmenev D, Wingender E. Databases and tools for in silico analysis of regulation of gene expression. In: Handbook of Toxicogenomics: Strategies and Applications. Edited by Jürgen Borlak, Copyright © 2005 WILEY-VCH Verlag GmbH & Co. KGaA, Weinheim, ISBN: 3-527-30342-1, pp. 253-290.
16. Wingender E, Chen X, Fricke E., Geffers R, Hehl R, Liebich I, Krull M, Matys V, Michael H., Ohnhäuser R, Prüß M, Schacherer F, Thiele S, Urbach S. The TRANSFAC system on gene expression regulation. *Nucleic Acids Res.* 2001; 29: 281-283.
17. Kolchanov NA, Podkolodnaya OA, Ananko EA, Ignatieva EV, Stepanenko IL, Kel-Margoulis OV, Kel AE, Merkulova TI, Goryachkovskaya TN, Busygina TV, Kolpakov FA, Podkolodny NL, Naumochkin AN, Korostishevskaya IM,

Romashchenko AG, Overton GC. Transcription Regulatory Regions Database (TRRD): its status in 2000 Nucleic Acids Res. 2000; 28:298-301

18. Dermitzakis ET, Reymond A, Scamuffa N, Ucla C, Kirkness E, Rossier C, Antonarakis SE. Evolutionary discrimination of mammalian conserved non-genic sequences (CNGs). *Science*. 2003; 302:1033-5.

19. Kel OV, Romaschenko AG, Kel AE, Wingender E, Kolchanov NA. A compilation of composite regulatory elements affecting gene transcription in vertebrates. *Nucleic Acids Res*. 1995; 23:4097-4103.

20. Kel-Margoulis OV, Ivanova TG, Wingender E, Kel AE. Automatic annotation of genomic regulatory sequences by searching for composite clusters. *Pac Symp Biocomput* 2002;:187-98.

21. Merika, M., and Thanos, D. Enhanceosomes. *Curr Opin Genet Dev*. 2001;11:205-208.

22. Eisen M, Spellman P, Brown P, Botstein, D. *Proc Natl Acad. Sci* USA 1998;95:14863-14868

23. Sabatti C. Statistical Issues in Microarray Analysis. Curr Genomics 2002;3:7-12

24. International Human Genome Sequencing Consortium. Initial sequencing and analysis of the human genome. *Nature*. 2001; 409:860-921.

25. Cawley S, Bekiranov S, Ng HH, Kapranov P, Sekinger EA, Kampa D, Piccolboni A, Sementchenko V, Cheng J, Williams AJ, Wheeler R, Wong B, Drenkow J, Yamanaka M, Patel S, Brubaker S, Tammana H, Helt G, Struhl K, Gingeras TR. Unbiased mapping of transcription factor binding sites along human chromosomes 21 and 22 points to widespread regulation of noncoding RNAs. *Cell* 2004; 116:499-509

26. Kel AE, Gossling E, Reuter I, Cheremushkin E, Kel-Margoulis OV, Wingender E. MATCH: A tool for searching transcription factor binding sites in DNA sequences. *Nucleic Acids Res*. 2003 Jul 1;31:3576-9.

27. Lawrence, CE, Altschul, SF, Bogouski MS, Liu JS, Neuwald AF, Wooten, JC. Detecting subtle sequence signals: a Gibbs sampling strategy for multiple alignment. *Science*, 262: 208-214.

28. Cheremushkin ES, Kel AE. Whole Genome Human/Mouse Phylogenetic Footprinting of Potential Transcription Regulatory Signals. *Pac Symp Biocomput*. 2003; 8:291-302.

29. Needleman SB, Wunsch CD. A general method applicable to the search for similarities in the amino acid sequence of two proteins. *J Mol Biol,* 1970; 48: 443-53

30. Kel A.E., Ponomarenko M.P., Likhachev E.A., Orlov Yu.L., Ischenko I.V., Milanesi L., Kolchanov N.A. SITEVIDEO: A computer System for Functional site Analysis and Recognition. Investigation of the human splice sites.(1993) *Comput. Applic. Biosci.* 9, 617-627.

31. Alexander E. Kel, Olga V. Kel-Margoulis, Peggy J. Farnham, Stephanie M.Bartley, Edgar Wingender and Michael Q. Zhang, Computer-assisted identification of cell-cycle related genes: New targets for E2F transcription factors. *J.Mol.Biol.* (2001) 309, 99-120.

32. Kel, A., Reymann, S., Matys, V., Nettesheim, P., Wingender, E., Borlak, J. A novel computational approach for the prediction of networked transcription factors of aryl hydrocarbon-receptor-regulated genes. *Mol. Pharmacol.* 66, 1557-1572 (2004).

Chapter 10

Web Servers for the Prediction of Curvature as well as other DNA Characteristics from Sequence

Kristian Vlahovicek[1], László Kaján[1], Gábor Szabó[2], Valentina Tosato[1],Carlo V. Bruschi[1], Sándor Pongor[1]

[1]International Centre for Genetic Engineering and Biotechnology (ICGEB), Area Science Park, Padriciano 99, 34012 Trieste, Italy

Contact: Sándor Pongor
Department of Biophysics and Cell Biology, University Medical School of Debrecen, 4012 Debrecen, agyerdei krt. 98, Hungary
Email: pongor@icgeb.org

10.1 Introduction

As very large amounts of genomic sequence data are generated, there is a growing need for simple methods that can guide experimenters to find regions that are conspicuous in terms of some structural properties, such as flexibility or intrinsic curvature. Over the past years we have been interested in developing and testing simple mechanic models that can describe the local behaviour of DNA in such short segments, in a sequence dependent fashion[1-9]. These methods have been incorporated into WWW-based server programs located on the ICGEB web site and the calculations have been extended to the calculation and visualization of various parameters other than curvature[10].

Generally speaking, parametric visualization consists in mapping of numerical data to visually presentable models. The simplest form of parametric visualization is the sequence plot i.e. a graph in which numeric values are assigned to positions along the DNA sequence. The advantage of comparing sequence plots rather than primary sequences originates from the simple fact that plots, unlike primary sequences, can be subjected to arithmetic operations (averaging, subtraction, etc.) and their similarities can be characterized in quantitative terms such as correlation coefficients and standard deviations. This is essentially a parametric approach of sequence comparison which makes it possible, e.g., to compare groups of sequences, to carry out a semiquantitative comparison (ranking) of sequences in structural terms, etc. using simple programs. The parametric visualization of DNA sequences uses the same properties on a qualitative basis, and the conspicuous segments can be identified by 1D, 2D or 3D plots of various parameters.

In this chapter we first describe DNA curvature as the paradigmatic concept, followed by a short description of the server algorithms. The last part of this chapter gives examples of applications.

10.2 Calculation of DNA curvature

The thinking of biologists has been profoundly influenced by the idea of local structural polymorphism in DNA. DNA is no longer considered as a featureless polymer but rather as a series of individual domains differing in flexibility and curvature. Unlike in the case of helical polymorphism (e.g. B, A or Z structures), here we often deal with a localised micropolymorphism in which the original B-DNA structure is only distorted but is not extensively modified[9]. The deviations from ideal, straight DNA are usually expressed as angles of deflection between adjacent basepairs (Figure 10-1A).

The terms "curved DNA" or "DNA curvature" are used in various contexts. For instance, asymmetrical binding of proteins can induce both kinks and smooth bends in the DNA trajectory. In this review we attempt to summarize another phenomenon, an inherent structural micro-heterogeneity of DNA that occurs in the absence of bound proteins, and depends only on the DNA sequence. In contrast to alternative DNA conformations (such as A and Z-DNA), curvature can be viewed as a slight distortion of the B-DNA geometry that is manifested in the bending of the

217

DNA-trajectory. Such a curvature can be quantitatively described using an analogy of a smoothly bent rod, and in the case of a DNA model, it can be expressed in terms of degree per base pair, or degree per helical turn. In the latter case, the repeat of the helical turn has to be specified. (**Figure 10-1B**).

Roll (ρ)　　Twist (Ω)　　Tilt (τ)

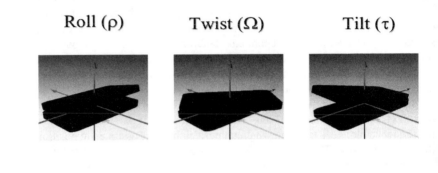

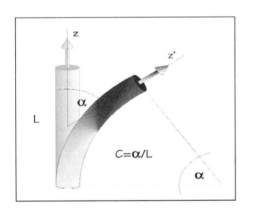

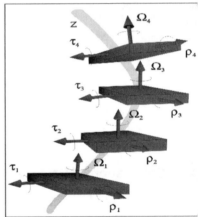

Figure 10-1. A.The molecular parameters describing DNA curvature are assigned to the relative orientation of two successive dinucleotides: roll angle (ρ), tilt angle (τ) and twist (Ω). In the ideal, Watson-Crick model, ρ=τ=0 and Ω=36° (10 basepairs per helical turn), B DNA in solution has a twist angle Ω=34.3° (10.5 basepairs per helical turn); for a detailed description of these and other parameters see[54]. B. Macroscopic curvature of an elastic rod is characterized by a deflection angle a, in the case of DNA this is sometimes expressed in degrees per helical turn. C. The experimentally determined conformation of DNA can be characterized by local roll, tilt and twist angles, and these values can be used to reconstruct the trajectory of the Z-axis.

The discovery of DNA curvature was a slow process. The first evidence that there is an influence of base composition on the average twist between adjacent base pairs came from DNA X-ray fiber diagrams, 20 years after the double-helix paper of Watson and Crick[11]. Subsequent studies by gel-electrophoresis[12], nucleotide/digestions[13] and finally the first X-ray structure of DNA[14] confirmed this view. In 1980, Trifonov and Sussman suggested a correlation between the helical repeat of the DNA and spacing of certain dinucleotides (especially AA and TT) along the sequence which indicated that a substantial part of eukaryotic DNA may in fact be curved[15]. Subsequent experimental data by Marini et al.[16] indicated that periodic A-tracts repeating in phase with the helical repeats cause curvature, which was confirmed both by electron microscopy[17] and by enzymatic circularization experiments[18]. By the mid nineties, the concept of DNA curvature became generally accepted, and even the apparent controversy between X-ray crystallography and solution experiments could be reconciled by the discovery that divalent cations induce a sequence dependent curvature in DNA[3].

A "curvature model" is a way to derive sequence-dependent DNA geometry parameters from experimental data. The models are different both in terms of the experimental data and the method of calculation. For example, it is common to fix some of the base-pair parameters at the values corresponding to straight B-DNA while letting others vary in a sequence-dependent fashion. In addition, the angles can be assigned to dinucleotides or to trinucleotides; these datasets are referred to as dinucleotide or trinucleotide scales. (All the models described here refer to double-stranded DNA molecules with "classic" phosphate orientations.)

The Wedge Model. The wedge model is called a "nearest neighbour model" since the geometry of a stack of two base pairs is considered to be defined by the two constituent nucleotides, and the influence of more distant neighbours is ignored[19]. The model is based on gel-electrophoresis data, described in terms of dinucleotide parameters, roll and tilt angles.

The Junction Model. The junction model was proposed based on gel-mobility experiments using oligonucleotides with "phased" (suitably spaced) adenine tracts[20, 21]. According to this model, curvature is caused by a deflection at each junction between the axes of the normal B-DNA and the B'-DNA of the poly dA, poly dT. The model assumes that the deflection at junction is a result of negative base-pair inclination in adenine tracts and zero inclination in the intervening B-DNA segments, and that this difference generates the bend[21]. According to Haran et al.[22] the wedge and the junction models are not necessarily incompatible. It appears, however, that there are events of curvature that neither the junction model nor the wedge model can sufficiently explain. For example, some GC-rich motifs, such as GGGCCC and CCCGGG have been showed opposite direction of bending[3] to those predicted from both models.

The Elastic Rod Model. The elastic rod model is based on DNAseI digestion data[1]. This enzyme bends the substrate towards the major grove, so the resulting model allows only one direction of bending, towards the roll angle. The original method described DNA bending in terms of a dimensionless parameter, "relative bending propensity" determined for trinucleotides[1, 4]. Subsequently, a physical model of sequence-dependent anisotropic-bendability (SDAB) was developed[9]. SDAB considers DNA to be an elastic rod, in which the flexibility of each segment (di- or trinucleotide) is anisotropic, namely, greater towards the major groove than it is in other directions. As DNAseI cannot distinguish between *a priori* bent and dynamically "bendable" sites, curvature according to this model is both static as well as dynamic in nature and can be recognized by the phased distribution of bent/bendable sites along the sequences. This can be visualized as a vectorial property along the sequence (Figure 10-2) which is conceptually analogous to the hydrophobic moment calculations in protein sequences.

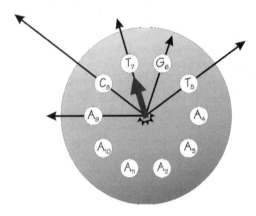

Figure 10-2. *DNA curvature as asymmetric bendability. The diagram is a top-view of the DNA helix with the Z-axis perpendicular to the plain of the paper. DNA bendability of subsequent trinucleotides is represented as an arrow perpendicular to the Z-axis. In curved segments, such as the one in the figure, the distribution of the bendability vectors is asymmetrical and the vector-sum (red arrow) is non-zero. In most parts of the genomes the vector sum is small[5].*

There are a number of computer programs that can predict curvature from sequence. One of the first algorithms available for curvature calculations was BEND written by Goodsell and Dickerson[23]. The algorithm can handle both dinucleotide and trinucleotide descriptions, and uses a simplified procedure wherein the successive deflection angles (roll, tilt) are summed up as vectors. This is a well-known approximation that is acceptable however only for low angle values. The BEND algorithm calculates curvature for segments of 11 nucleotides, and outputs a plot of curvature versus sequence position. The algorithm was incorporated into the EMBOSS suite of sequence analysis programs[24] under the name BANANA (which is a reference to curved B-DNA of A and non-A tracts), and is also available on-line[25]. The Haifa University server[26] for DNA structure calculation is built around the program Curvature[27]. The DIAMOD program was written by Mensur

Dlakic for PC[28] and handles most curvature models. Finally, several precomputed parametric genome maps are available in the Genome Atlas of the Technical University of Denmark[29, 30].

10.3 Prediction of DNA properties other than curvature

From the computational point of view curvature is a local property of DNA that can be represented by numeric values assigned to each position of a DNA sequence. The same philosophy can be extended to a large number of other DNA properties that can be assigned to a short segment of DNA. There are a few common approximations underlying many parametric descriptions: a) The property is local, i.e. a given n-mer in DNA will have the same property irrespective of its sequence environment ("context"). This may be true for molecular properties depending only on the nucleobases, but is a very rough approximation for complex, statistically derived properties like conformational preferences since, for instance, even dinucleotides are known to adopt a few different conformations depending on their neighbours. b) Segments within DNA (nucleotides, dinucleotides) contribute independently to a given property. This makes it possible to use simple linear or log/linear models to experimental data.

As an example, bending propensity parameters for trinucleotides were deduced from DNAseI digestibility vs. sequence data based on the following principles[1]. (i) Locality: DNase I interacts with the window of 6 nucleotides around the cleaved bond and its cutting efficiency depends only on this window. (ii) This window is represented as four overlapping trinucleotides, and one single structural parameter $p(a)$ of the trinucleotides, constituting the enzyme-DNA contact surface, will influence the cutting rate (this is an obvious simplification, since local effects, such as specific residue contacts between the enzyme and the DNA molecule, are not considered); (iii) the bending propensity $p(a)$ of each trinucleotide contributes independently to the probability of DNase I cutting, P_W. The model thus assumes that the contribution of one element (trinucleotide) does not depend on any other element being present or absent in the window around the cut. So P_W for the 6 nt window can be written as the product of the n different and assumedly independent $p(a)$ probabilities:

(1)

$$P_w = \prod_1^4 p(a)_i$$

Equating PW with the experimentally determined frequencies of cleavage, FW, leads to a linear system of equations

(2)

$$F_w = \sum_1^4 \ln p(a)_i$$

221

Similar approaches have been used to extract numeric parameters from a wide variety of different experimental data. As an extreme case, DNAseI digestibility data can be obtained on large, continuous DNA fragments, other parameters, such as stability etc. were derived from measurements on short oligonucleotides. Regarding the origins of the data, parameters can be obtained either from measurement or from database statistics, such as evaluation of 3D structures or sequence data. From the computational point of view, the parameters are represented either as tabulated values, or they are computed "on the fly", based on the sequence information itself.

10.4 The DNA-analysis tools developed at ICGEB

The plot.it server produces parametric plots using various statistical physicochemical parameters[31]. A query sequence is divided into overlapping n-mers, and the average value of a given parameter is calculated using tabulated values. The server uses 45 structural parameters (a full list of references is available at the site), the general scheme of calculations is shown in Figure 10-3. The results appear either as simple sequence plots or as 2-D plots in which two parameters are plotted against each other. Examples are shown in Figure 10-4.

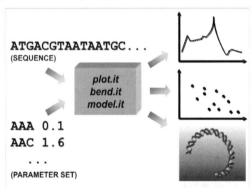

Figure 10-3. Data flow of the bend.it, plot.it and model.it servers. Each overlapping triplet (or dinucleotide) in a DNA sequence is assigned a corresponding parameter value in a "sliding window" fashion. The resulting numerical vector can then be averaged within a given window (default value is 31bp or approximately three helical turns) and displayed either as a 1C parameter vs. sequence plot, or as a 2D correlation plot from two different parameter sets. (Three-dimensional DNA trajectories are built from basepair geometry parameters without averaging).

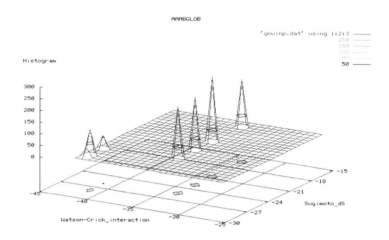

Figure 10-4. *Output examples of the plot.it server. A three-dimensional correlation plot of the Anadara trapezia (ark clam) beta globin gene (complete cds. genbank:L16978). The vertical Z-axis denotes the number of actual segments represented by data on the XY plane. This type of correlation plot is useful in situations where analysis is performed on a long DNA sequence.*

The bend.it server calculates the curvature of DNA molecules as predicted from the DNA sequences. The calculation is based on values tabulated for dinucleotides and trinucleotides, and the curvature (degree per helical turn) is calculated using standard algorithms[9]. This calculation was originally based on DNA bendability parameters derived from DNAseI digestion that characterize the (static or dynamic) bending of trinucleotides towards the major groove[2]. Today a number of other dinucleotide[32-34] and trinucleotide models[2, 4] are included, and the results can be visualized as 1D or 2D plots on the screen. Both the bend.it and the plot.it servers are based on C programs provided with GnuPlot graphic routines. Their output appears on the screen and is optionally sent by e-mail to the user (Figure 10-5).

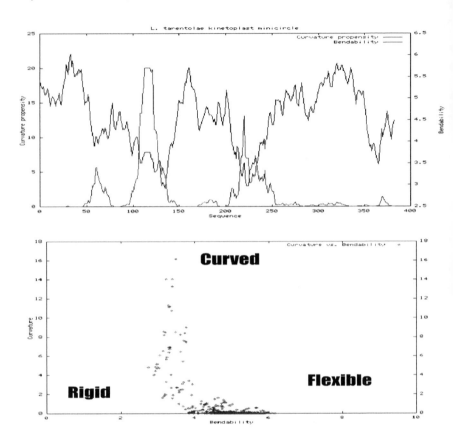

Figure 10-5. *Output examples of the bend.it server. A. profile plots of bendability (black) and curvature (gray) along the 350bp L. tarantolae kinetoplast sequence. Profile plots provide a visual aid to locate "interesting" regions along a DNA sequence. B. correlation (2D) plot of curvature vs. bendability of the same sequence.*

The model.it server was designed to provide 3D models of DNA in response to DNA sequence queries[35]. The results are presented as a standard PDB file that can be viewed directly using any of the widely available molecule manipulation programs such as Swiss-PDBviewer[36] or Rasmol[37]. In addition to straight A and B DNA models, the server is capable of building curved DNA models using the parameter sets mentioned above. The server program was written using "NAB" — a high level molecule manipulation language[38]. Coordinates of the sugar-phosphate backbone are optionally optimised with constrained molecular dynamics using energy parameters from the AMBER package[39]. At present, the server can produce models of 700 bp in length, but models longer than 50 bp will not be optimised. Modelling of canonical, straight B or A DNA structures proceeds in a similar way, but without the need for backbone geometry optimisation (Figure 10-6).

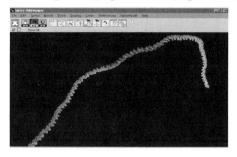

Figure 10-6. Output examples of the model.it server[35]. A. three-dimensional trajectory model of a ~400bp L. tarantolae kinetoplast, visualized using SwissPDB Viewer. B. Predicted conformation of 14 Zea mays promoter regions from EPD database ORF is shown in white. C. Superposition reveals three conformation groups.

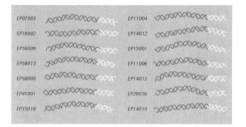

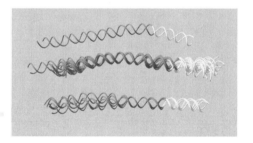

The IS introns server was designed to provide statistical overviews on intron groups[40]. Simple questions, like comparison of introns between various taxonomic groups in terms of intron phases or size-distributions as well as the analysis of splice sites, requires a carefully selected dataset as well as meticulous work that has to be repeated as new data become available. The goal of the introns server was to establish an automatically updated intron resource that allows the evaluation of experimentally validated and statistically balanced intron datasets, as well as a flexible comparison of groups according to various criteria. In addition to sequence retrieval and BLAST similarity search, there are options to compare taxonomic groups based on the NCBI Taxonomy Database, and to perform on the fly statistics. The analysis capabilities include statistical evaluation (minimum, maximum, average, standard deviation, etc.)

of intron and exon length, of the number of introns per gene, base composition, intron phases, as well as a graphic comparison of two or more groups in terms of the above variables. In addition, the analysis of splice sites and testing of the exon shuffling hypothesis[41, 42] are explicitly included (Figure 10-7).

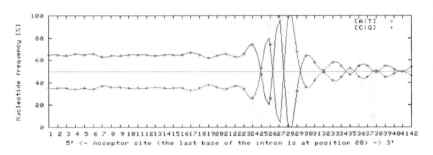

Observed and expected number of intron phase associations

N of exons	Observed / expected number									Sample size	ChP
	0..0	1..1	2..2	0 1	0 2	1 0	1 2	2 0	2 1		
1:	998/903	161/163	173/163	344/388	380/388	401/388	141/163	370/388	158/163	3126	16.3 +
2:	920/839	138/122	138/127	255/302	318/325	316/340	109/121	314/327	123/118	2631	24.5 +++
3:	805/730	95/107	105/110	248/263	253/282	253/286	128/115	295/285	107/103	2289	21.3 ++
4:	715/637	91/93	96/96	202/229	225/247	217/250	105/100	261/240	66/69	1998	22.0 +++
5:	613/554	79/81	86/84	170/200	216/215	197/225	73/67	225/216	68/70	1739	19.6 +
6:	551/478	65/70	68/72	130/172	192/185	158/194	77/75	193/186	61/67	1500	30.3 +++
7:	502/414	57/61	76/63	107/149	143/160	133/168	70/85	151/162	59/58	1300	44.6 +++

*Number of exons incorporated in the intron phase association.
**Confidence levels + = 95%, ++ = 99%, +++ = 99.5%.

Figure 10-7. Output examples of the intron server. A. Nucleotide frequency plot, characteristic for plant (taxonomic group: Viridiplantae) introns near the acceptor site.Gray: adenine/thymin positional frequency.Black: cytidine/guanine positional frequency. B. Intron phase association statistics for plants given in terms of observed and expected intron phase associations. "0..0" (column) denotes a set of introns string with phase 0 and finishing with a phase 0. The number of intercalated exons is shown in the first column, the other possible intron phase associations are given in separate columns. The significance is shown by the high Chi-square values, the level of significance is 95% as opposed to 99.5% found for animal introns. This supports the hypothesis that exon shuffling is a probable scenario in plant evolution.

All the servers are provided with help files that describe the detailed instructions, the theory, the literature citations as well as the instructions for installing the accessory programs such as Swiss-PDBviewer[36] or Rasmol[37].

10.5 Application examples

One of the obvious applications is to compare the distribution of curvature and other parameters in genomic sequences. Figure 10-8 shows that bendability has a smooth, symmetrical distribution in genomic DNA, similar to a bell shape. The distribution of curvature is apparently non-symmetrical reminiscent of a gamma function which is often found with randomly distributed variables whose value cannot be negative - curvature is actually such a case. Another possibility is to analyse curved segments along the entire genome. A circular plot is a convenient way to show such distributions even though the graphic resolution is often a limiting factor. Another possibility is to analyse the vicinity of annotated features in genomes, as shown in Figure 10-9.

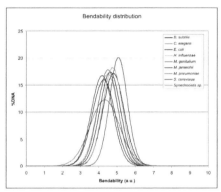 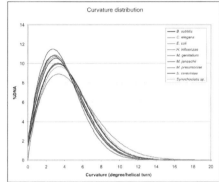

Figure 10-8. *Distribution of bendability and curvature in various prokaryotic genomes.*

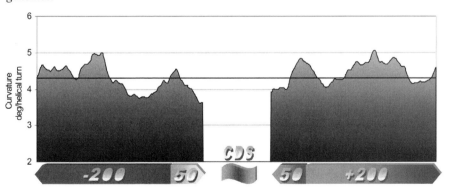

Figure 10-9. *Distribution of curvature around open reading frames in yeast chromosome III. Curvature profiles of all fragments 200 bp outside and 50 bp inside the open reading frame were averaged and the result is displayed in terms of a positional preference for curved regions with respect to start and stop codons. The black line represents the average curvature of yeast chromosome III.*

227

A comprehensive analysis of the curvature of the *B. subtilis* genomic DNA revealed the percentage of curved motifs within the genome and how many ORFs contain curved segments[43]. As reported in Figure 10-10, less than 1% of the *B. subtilis* genome contains curved motifs with values above 14° per helical turn. Using this as a cut-off, the majority of the curved DNA is found within the ORFs while using 16°, 64% of the curved segments are within the intergenic regions (Figure 10-10, inset), a tendency that continues as the cut-off is raised. In other words, the majority of the most curved segments are concentrated in the intergenic regions.

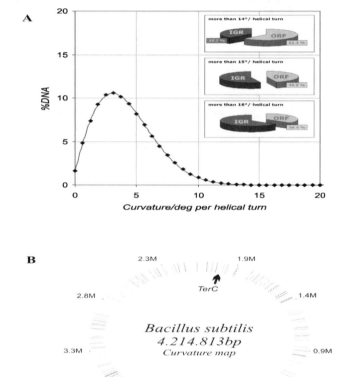

Figure 10-10. *Distribution of the curved segments within the B. subtilis genome. The distribution of the curved motifs inside Open Reading Frames (ORFs) or inside InterGenic Regions (IGR) obtained with different cut-off degree values is represented in the inset. The graphic representation of the curved motifs is reported at the bottom.*

Figure 10-11 shows the number of ORFs with at least one curved motif. Only 6.2% of all the ORFs in *B. subtilis* shows a curvature with a cut-off value of 14°. These ORFs with at least one curved motif, encode functionally unrelated proteins since their percentage distribution is consistent with the distribution of the known proteins among the different classes, following the standard functional classification reported by SubtiList (cellular processes and cell envelope, intermediary metabolism, information pathways, other functions,[44]). Therefore, only a small percentage of all curved motifs fall inside the coding regions, leading to the hypothesis that a straight DNA is more efficiently transcribed. On the other hand, it could be that intergenic regions have been selected with an intrinsic high curvature to act as genomic signals. Indeed, it is known that, at least in lower eukaryotes such as *Saccharomyces cerevisiae* and *Leishmania major*, promoters and terminators are constituted by flexible DNA stretches[45]. Several coding strand switching points are present within the chromosomes of *L. major*[46]. For example in a region of 74 kb of chromosome 21, the first 16 ORFs are encoded on the Crick strand, while the rest of them are localised on the Watson strand (Figure 10-12). Between the two coding regions there are 1,602 nt, which are part of the so-called switching region, that do not contain neither predicted CDS nor DNA with potential to form hairpin structures. Moreover, this region shows a high DNA curvature with a maximum value of GC skew, as detected by the Bend-it program[46, 47].

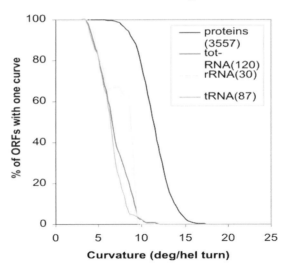

Curvature in the ORFs of the *B. subtilis* genome

Figure 10-11. Percentage of ORFs coding for proteins, RNA, rRNA and tRNA, overlapping with at least one curved motif.

229

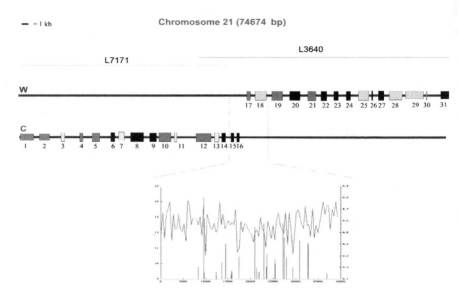

Figure 10-12. *Region of 74 kb around the switching point of chromosome 21 of L. major. L7171 and L3640 are two cosmids containing the overlapping fragments of chromosome 21. The two strands and the encoded genes are represented in different shadows of gray. The curvature analysis of the 40 kb around the switch region is shown in the lower window.*

The physical features described for the switching point of chromosome 21 characterise also the switching points of other chromosomes of the parasite (chromosomes 1, 3, 4, 19), suggesting that these regions can be involved in promotion of DNA transcription or can indicate the presence of an origin of replication. In support of the first hypothesis, very recently it was shown, by transfectional studies, that the switching point region seems to drive the expression of the entire chr1 in *Leishmania major Friedlin*[48].

DNA isolated from normal eukaryotic cells by standard methods exhibit particular fragility resulting in ~50 kb fragments. Breakage at hypersensitive/fragile sites is thought to be due to nucleolytic cleavage and/or localized, non-random release of torsional stress[49-52]. The sequence of several breakpoints of human DNA was recently determined and by multiple alignement, sequence similarities were found among the various breakpoints, both in short and longer stretches of the DNA[53]. An analysis with the *plot.it* server showed peculiar conformational characteristics (sharp transition or with a center of symmetry) located exactly at the experimentally determined breakpoints (Figure 10-13A)[53]. These, however, did not exactly coincide with the position of the short consensus motives. A number of short consensus motives appear to have a curved conformation as predicted by the model.it server (Figure 10-13B). These instances of correlation between computed and biochemical behavior imply that the predicted conformations may be useful in the analysis situations where breakage and rearrangements are implicated in pathological scenarios.

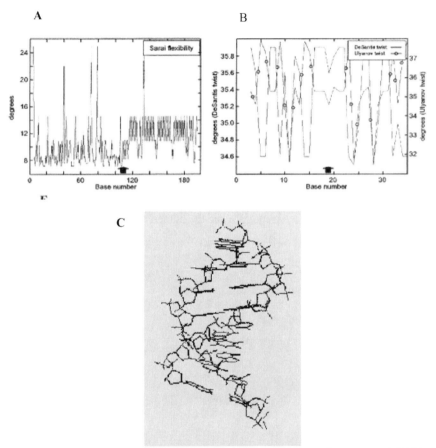

Figure 10-13. *Analysis of breakpoint sequences in human chromosomal DNA A: Flexibility of DNA obtained from conformational energy calculations[55], expressed as dinucleotide twist, roll and tilt angles, in the case of breakpoint sequence no 10 (numbering given in [53]) Thick arrow: breakpoint. (plot.it server) B. Twist angles in sequence no. 1 (40 bp) determined from NMR data (empty circles;[32]), and as predicted based on conformational energy calculations[56]. Thick arrow: breakpoint. (plot.it server) C. Predicted 3-D model of the short breakpoint motif CCAGCCTGG, built by the model.it server using the consensus scale of DNA curvature[9], and the raw models being refined by simulated annealing[35] (model.it server).*

10.6 Summary

The WWW servers at ICGEB[10] have been created for the analysis of user-submitted DNA sequences in structural terms. *bend.it* calculates DNA curvature according to various methods, *plot.it* creates parametric plots of 45 physicochemical as well as statistical parameters. Both programs provide 1D as well as 2D plots that allow localisation of peculiar segments within the query. *model.it* creates 3D models of canonical or bent DNA starting from sequence data and presents the results in the form of a standard PDB file, directly viewable on the user's PC using any molecule manipulation program. The *introns* server allows statistical evaluation of introns in various taxonomic groups and the comparison of taxonomic groups in terms of length, base composition, intron type etc. The options include the analysis of splice sites and a probability test for exon-shuffling.

The application examples cited here show that in some cases, genomic segments identified by parametric analysis show interesting correlations even in the absence of sequence similarity. However the correlation is generally weak, so careful analysis and human experts are necessary for the evaluation of the results. On the other hand, parametric plots can be excellent subjects for machine learning studies that might in turn reveal correlations that currently escape the human eye.

References

1. Brukner, I., et al., Sequence-dependent bending propensity of DNA as revealed by DNase I: parameters for trinucleotides. *Embo J.*, 1995. 14(8): p. 1812-8.
2. Brukner, I., et al., Trinucleotide models for DNA bending propensity: comparison of models based on DNaseI digestion and nucleosome packaging data. *J. Biomol. Struct. Dyn.*, 1995. 13(2): p. 309-17.
3. Brukner, I., et al., Physiological concentration of magnesium ions induces a strong macroscopic c urvature in GGGCCC-containing DNA. *J. Mol. Biol.*, 1994. 236(1): p. 26-32.
4. Gabrielian, A. and S. Pongor, Correlation of intrinsic DNA curvature with DNA property periodicity. *Febs Lett*, 1996. 393(1): p. 65-8.
5. Gabrielian, A., A. Simoncsits, and S. Pongor, Distribution of bending propensity in DNA sequences. *Febs Lett*, 1996. 393(1): p. 124-30.
6. Gabrielian, A., K. Vlahovicek, and S. Pongor, Distribution of sequence-dependent curvature in genomic DNA sequences. *FEBS Letters*, 1997. 406(1-2): p. 69-74.
7. Gromiha, M.M., et al., Anisotropic elastic bending models of DNA. *J. Biol. Phys.*, 1996. 22: p. 227-243.
8. Gromiha, M.M., et al., The role of DNA bending in Cro protein-DNA interactions. *Biophys. Chem.*, 1997. 69(2-3): p. 153-60.
9. Munteanu, M.G., et al., Rod models of DNA: sequence-dependent anisotropic elastic modelling of local bending phenomena. *Trends Biochem. Sci.*, 1998. 23(9): p. 341-7.
10. www.icgeb.org/dna.
11. Bram, S., Variation of type-B DNA x-ray fiber diagrams with base composition. *Proc. Natl. Acad. Sci. U.S.A.*, 1973. 70(7): p. 2167-70.
12. Wang, J.C., Helical repeat of DNA in solution. *Proc. Natl. Acad. Sci. U.S.A.*, 1979. 76(1): p. 200-3.
13. Dickerson, R.E. and H.R. Drew, Kinematic model for B-DNA. *Proc. Natl. Acad. Sci. U.S.A.*, 1981. 78(12): p. 7318-22.
14. Dickerson, R.E. and H.R. Drew, Structure of a B-DNA dodecamer. II. Influence of base sequence on helix structure. *J. Mol. Biol.*, 1981. 149(4): p. 761-86.
15. Trifonov, E.N. and J.L. Sussman, The pitch of chromatin DNA is reflected in its nucleotide sequence. *Proc. Natl. Acad. Sci. U.S.A.*, 1980. 77(7): p. 3816-20.
16. Marini, J.C., et al., A bent helix in kinetoplast DNA. *Cold Spring Harb. Symp. Quant. Biol.*, 1983. 47 Pt 1: p. 279-83.
17. Griffith, J., et al., Visualization of the bent helix in kinetoplast DNA by electron microscopy. *Cell*, 1986. 46(5): p. 717-24.
18. Ulanovsky, L., et al., Curved DNA: design, synthesis, and circularization. *Proc. Natl. Acad. Sci. U.S.A.*, 1986. 83(4): p. 862-6.
19. Ulanovsky, L.E. and E.N. Trifonov, Estimation of wedge components in curved DNA. *Nature*, 1987. 326(6114): p. 720-2.
20. Diekmann, S., Sequence specificity of curved DNA. *Febs Lett*, 1986. 195(1-2):

p. 53-6.

21. Koo, H.S. and D.M. Crothers, Calibration of DNA curvature and a unified description of sequence-directed bending. *Proc. Natl. Acad. Sci. U.S.A.*, 1988. 85(6): p. 1763-7.

22. Haran, T.E., J.D. Kahn, and D.M. Crothers, Sequence elements responsible for DNA curvature. *J. Mol. Biol.*, 1994. 244(2): p. 135-43.

23. Goodsell, D.S. and R.E. Dickerson, Bending and curvature calculations in B-DNA. *Nucleic Acids Res.*, 1994. 22(24): p. 5497-503.

24. http://www.hgmp.mrc.ac.uk/Software/EMBOSS/.

25. http://www.hgmp.mrc.ac.uk/Software/EMBOSS/interfaces.html.

26. http://esti.haifa.ac.il/~leon/cgi-bin/curvatur/.

27. Shpigelman, E.S., E.N. Trifonov, and A. Bolshoy, CURVATURE: software for the analysis of curved DNA. *Comput. Appl. Biosci.*, 1993. 9(4): p. 435-40.

28. http://www-personal.umich.edu/~mensur/software.html.

29. Pedersen, A.G., et al., A DNA structural atlas for Escherichia coli. *J. Mol. Biol.*, 2000. 299(4): p. 907-30.

30. Jensen, L.J., C. Friis, and D.W. Ussery, Three views of microbial genomes. *Res. Microbiol.*, 1999. 150(9-10): p. 773-7.

31. Vlahovicek, K., A. Gabrielian, and S. Pongor, Prediction of bendability and curvature in genomic DNA. *J. Mathematical Modelling and Scientific Computing*, 1998. 9: p. 53-57.

32. Ulyanov, N.B. and T.L. James, Statistical analysis of DNA duplex structural features. *Methods Enzymol.*, 1995. 261(120): p. 90-120.

33. Bolshoy, A., et al., Curved DNA without A-A: experimental estimation of all 16 DNA wedge angles. *Proc. Natl. Acad. Sci. U.S.A.*, 1991. 88(6): p. 2312-6.

34. Olson, W.K., et al., Influence of fluctuations on DNA curvature. A comparison of flexible and static wedge models of intrinsically bent DNA. *J. Mol. Biol.*, 1993. 232(2): p. 530-54.

35. Vlahovicek, K. and S. Pongor, Model.it: building three dimensional DNA models from sequence data. *Bioinformatics*, 2000. 16(11): p. 1044-5.

36. Guex, N. and M.C. Peitsch, SWISS-MODEL and the Swiss-PdbViewer: an environment for comparative protein modeling. *Electrophoresis*, 1997. 18(15): p. 2714-23.

37. Sayle, R.A. and E.J. Milner-White, RASMOL: biomolecular graphics for all. *Trends Biochem. Sci.*, 1995. 20(9): p. 374.

38. Macke, T. and D.A. Case, Modeling unusual nucleic acid structures, in Molecular Modeling of Nucleic Acids, N.B. Leontes and J. SantaLucia, Editors. 1998, *American Chemical Society*: Washington DC. p. 379-393.

39. Case, D.A., et al., AMBER 5. 1997, University of California: San Francisco.

40. Barta, E., L. Kajan, and S. Pongor, IS: A web-site for introns statistics. *Bioinformatics*, 2003. 19: p. 543.

41. Long, M., S.J. de Souza, and W. Gilbert, Evolution of the intron-exon structure of eukaryotic genes. *Curr. Opin. Genet. Dev.*, 1995. 5(6): p. 774-8.

42. Kriventseva, E.V. and M.S. Gelfand, Statistical analysis of the exon-intron

structure of higher and lower eukaryote genes. *J. Biomol. Struct. Dyn.*, 1999. 17(2): p. 281-8.

43. Tosato, V., et al., The DNA secondary structure of the Bacillus subtilis genome. *FEMS Microbiol Lett*, 2003. 218(1): p. 23-30.

44. http://bioweb.pasteur.fr/GenoList/SubtiList.

45. McDonagh, P.D., P.J. Myler, and K. Stuart, The unusual gene organization of Leishmania major chromosome 1 may reflect novel transcription processes. *Nucleic Acids Res.*, 2000. 28(14): p. 2800-3.

46. Tosato, V., et al., Secondary DNA structure analysis of the coding strand switch regions of five Leishmania major Friedlin chromosomes. *Curr. Genet.*, 2001. 40(3): p. 186-94.

47. Myler, P.J., et al., Genomic organization and gene function in Leishmania. *Biochem. Soc. Trans.*, 2000. 28(5): p. 527-31.

48. Martinez-Calvillo, S., et al., Transcription of Leishmania major Friedlin chromosome 1 initiates in both directions within a single region. *Mol. Cell*, 2003. 11(5): p. 1291-9.

49. Szabo, G., Jr., F. Boldog, and N. Wikonkal, Disassembly of chromatin into approximately equal to 50 kb units by detergent. *Biochem. Biophys. Res. Commun.*, 1990. 169(2): p. 706-12.

50. Szabo, G., Jr., 50-kb chromatin fragmentation in the absence of apoptosis. *Exp. Cell. Res.*, 1995. 221(2): p. 320-5.

51. Gal, I., et al., Protease-elicited TUNEL positivity of non-apoptotic fixed cells. J *Histochem. Cytochem.*, 2000. 48(7): p. 963-70.

52. Varga, T., I. Szilagyi, and G. Szabo, Jr., Single-strand breaks in agarose-embedded chromatin of nonapoptotic cells. *Biochem. Biophys. Res. Commun.*, 1999. 264(2): p. 388-94.

53. Szilagyi, I., et al., Non-random features of loop-size chromatin fragmentation. *J. Cell Biochem.*, 2003. 89(6): p. 1193-205.

54. Bansal, M., D. Bhattacharyya, and S. Vijaylakshmi, NUVIEW: software for display and interactive manipulation of nucleic acid models. *Comput. Appl. Biosci.*, 1995. 11(3): p. 289-92.

55. Sarai, A., et al., Sequence dependence of DNA conformational flexibility. *Biochemistry*, 1989. 28(19): p. 7842-9.

56. De Santis, P., et al., Validity of the nearest-neighbor approximation in the evaluation of the electrophoretic manifestations of DNA curvature. *Biochemistry*, 1990. 29(39): p. 9269-73.

235

Chapter 11

AlleleID — Pathogen Identification System

Arun Apte, Deepanjali Thakur

Contact:
Arun Apte
Chief Technology Officer
PREMIER Biosoft International
786 Corina Way
Palo Alto, CA 94303-4504 USA
Email : aapte@PremierBiosoft.com

11.1 Introduction

Pathogenic agent detection and control are one of the primary focuses of the pharmaceutical and biological research industries. Research in these areas has lead to the containment of previously common agents such as smallpox. Researchers and clinicians have long been aware of diseases caused by these pathogen and developed various techniques to identify these organism. AlleleID is a complete system for the design of real time assay and microarray identification of pathogens. There are two categories of pathogen detection assays, conventional and molecular. Conventional methods have longer turnaround time and, in many cases, lower sensitivity. Sequence-based molecular methods such as real time PCR, microarrays, and band biosensors provide high sensitivity, rapid diagnostics and higher specificity allowing differentiation between related strains. Of these methods, real time PCR is both sensitive and specific for pathogen detection. It is performed in sealed wells, which reduces the risk of cross-contamination, and does not require any post-PCR analysis. Real time PCR assays require primers and probes for pathogen identification. Manual design of these probes is both time consuming and results in lower quality probes due to the inability to simultaneously balance multiple criteria. It is frequently desirable to design cross species probes that will test positive for the same gene on multiple species. This enables a single probe to be used to study a gene using multiple species, thus resulting in significant cost savings. To test positive on multiple species, primers and probes should be designed in the conserved region of a gene. Manual identification of such conserved regions and design of optimal or standard PCR primers and probes for amplification and detection is a cumbersome job. A similar problem is encountered while the design of assays for organisms with close phylogeny eg. group or taxa. AlleleID is able to design both cross species and taxa specific pathogen identification probes. The system uses ClustalW application for multiple sequence alignment. ClustalW produces biologically meaningful alignments of divergent sequences, calculates the best match for the selected sequences and lines them up in such a manner that the identities, similarities and differences can be distinguished easily. The alignment allows easy identification of the conserved regions even among distantly related sequences. The alignment also highlights the regions of difference between groups of related sequences. The identified regions are used for designing species specific, taxa specific and cross species probe sets. These probes can then be used in various high throughput applications associated with pathogen detection such as real time PCR or DNA microarray assays.

11.2 Methods of Pathogen detection

11.2.1 Conventional Methods

Conventional methods for the detection of pathogens start with the culturing of bacteria onto agar plates. Culturing methods are time consuming requiring three to even days to determine a total viable count and to detect specific pathogenic bacteria. There are several disadvantages to conventional methods :

- Not all microbes can be detected by these methods. Eg : Salmonella

- Every microorganism requires some factor for energy generation and cellula biosynthesis. For example, chemoautotrophs or lithotrophs require inorganic compounds such as H_2, NH_3, NO_2, H_2S, and CO_2 for their growth. With conventional methods only pathogens with known growth requirements can be detected.

- Pathogens within common behavioral features such as having similar colony or staining characteristics when cultivated are poorly discriminated.

- Pathogens such as retroviruses which fail to elicit a detectable host immune response cannot be detected adequately.

11.2.2 Immunological Methods

Clinical, agricultural and environmental samples infected with pathogens are commonly tested using immunological methods[1]. Accuracy of the immunological methods depends on the affinity and specificity of the antibody used. High affinity antibodies are essential for the sensitivity required. Many immunonological method employ antibodies that react with components located on the surface of the pathogen e.g. virus capsid protein, cell wall or flagellar antigens. The most common technique used to detect pathogens immunologically are:

11.2.2.1 Electroimmunoassay

A specific antigen and antibody are used for production of an electrical signal. A circuit is prepared, which allows the attachment of a capture antibody on a solid surface near the electrode gap. Upon addition of the sample, the target antigen binds to the capture antibody and then, a colloidal gold–labeled detection antibody is bound which leads to the formation of a capture-target-detector sandwich. In the final step silver ions are deposited onto the colloidal gold which results into a conductive silver bridge, resulting in a drop in resistance because of the closing of the circuit.

11.2.2.2 ELISA

Enzyme linked immunosorbent assay (ELISA) techniques have the inherent ability to detect minute concentrations of pathogens, with no need for culturing Detection is based on antigen-antibody interaction, resulting in a color change. In the presence of the antigen, an enzyme conjugated with an antibody reacts with a colorless substrate, the chromogenic substrate, resulting in a colored reaction A number of variations of ELISA have been developed including indirect ELISA sandwich ELISA and competitive ELISA.

11.2.2.3 Immunofluorescence

This method is based on the same principle as ELISA and electroimmunoassay. Microscope slides are used on which cells, tissue, or some other substances infected with the pathogen are placed. Next, a small amount of serum containing antibodies is placed over the cells or tissue thus allowing the binding of the antibodies that are specific for a particular tissue or cellular antigens. The serum is washed away, and a second antibody that binds to human antibodies is applied to the slide. This second antibody has a fluorescent dye chemically linked to it. A bright fluorescence is observed when the antibodies of infected serum bind to the tissue or cell samples.

Disadvantages of Using Immunological Techniques

*   For detecting various strains of pathogens, highly specific antigenic determinants or epitopes are required and finding such determinants accurately is a difficult task

*   False positives occur when the antibody binds to alternate targets closely resembling the antigen. In these cases, sensitivity of the method can be affected.

11.2.3 Molecular Methods

Molecular methods including the polymerase chain reaction, real time quantitative PCR, and DNA microarrays have revolutionized the detection of pathogenic microorganisms. PCR variations used commonly in diagnostics are nested PCR, random amplified polymorphic DNA (RAPD), reverse transcriptase-PCR (RT-PCR), and reverse cross blot PCR[2]. By providing levels of sensitivity and specificity these molecular methods allow rapid detection of pathogens not possible with traditional culture-based assays. Molecular methods which are commonly used for pathogen detection are :

11.2.3.1 Quantitative PCR

The quantitative polymerase chain reaction (qPCR) is the most sensitive and reliable method for the detection and quantitation of nucleic acids. The primary advantage of real time PCR is that it allows the amplification process to be monitored during each cycle of the PCR process, rather than following completion as observed in general PCR reactions. This decreases the time required for detection and also enables quantification of starting target DNA. Specialized tagged probes such as molecular beacons[3] and TaqMan®[4] are used to increase specificity and to determine the purity of a PCR product:

11.2.3.1.1 Molecular Beacons

Molecular beacons are single stranded hairpin shaped structured oligonucleotide probes. In the presence of the target sequence, they unfold, bind and fluoresce. The molecule consists of a loop, stem, 5' fluorophore and 3' quencher moiety. Molecular beacon probes can report the presence of specific nucleic acids from both homogenous and solution. In the presence of a complementary target, the "stem" portion of the beacon separates out resulting in the probe hybridizing to the target. In the absence of a complementary target sequence, the beacon remains closed and there is no appreciable fluorescence. When the beacon unfolds in the presence of the complementary target sequence, the fluorophore is no longer quenched, and the molecular beacon fluoresces (Figure 11-1). The fluorescence is easily detected in a thermal cycler. The amount of fluorescence at any given cycle, or following cycling, depends on the amount of specific product. For quantitative PCR, molecular beacons bind to the amplified target following each cycle of amplification and the resulting signal is proportional to the amount of template. In addition, the stem probe structure of a molecular beacon makes it better able to discriminate single base-pair mismatches (compared to linear probes) because the hairpin makes mismatched hybrids thermally less stable than hybrids between the corresponding linear probes and their mismatched target. Furthermore, unlike linear hydrolysis probes, quenching of molecular beacons has been shown to occur through a direct transfer of energy from fluorophore to quencher. Consequently, a common quencher molecule can be used, increasing the number of possible fluorophores that can easily be used as reporters. This is an important advantage when designing PCR experiments in which several molecular beacons with different colored fluorophores are used to detect multiple targets in the same tube for multiplexing[5].

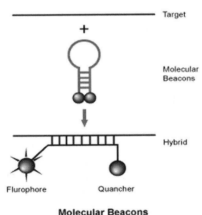

Molecular Beacons

Figure 11-1. (NOTE: Quancher should read Quencher)

11.2.3.1.2 TaqMan® Probes

TaqMan® probes are dual labeled hydrolysis probes which utilize the 5' ->3'exonuclease activity of the enzyme Taq Polymerase for measuring the amount of target sequences in the samples.TaqMan® probes consist of a 18-22 bp oligonucleotide probe which is labeled with a reporter fluorophore at the 5' end and a quencher fluorophore at the 3' end. While carrying out a TaqMan® experiment, a fluorogenic probe, complementary to the target sequence is added to the PCR reaction mixture (Figure 11-2). This probe is an oligonucleotide with a reporter dye attached to the 5' end and a quencher dye attached to the 3' end. Till the time the probe is not hydrolyzed, the quencher and the fluorophore remain in proximity to each other, separated only by the length of the probe. This proximity however, does not completely quench the fluorescence of the reporter dye and a background flourescence is observed. During PCR, the probe anneals specifically between the forward and reverse primer to an internal region of the PCR product. The polymerase then carries out the extension of the primer and replicates the template to which the TaqMan® is bound. The 5' exonuclease activity of the polymerase cleaves the probe, releasing the reporter molecule away from the close vicinity of the quencher. The fluorescence intensity of the reporter dye, as a result increases. This process repeats in every cycle and does not interfere with the accumulation of PCR product. TaqMan® probes are used in multiplex real time reaction, allele specific assays and for pathogen identification as well[6].

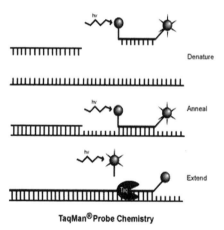

TaqMan®Probe Chemistry

Figure 11-2.

11.2.3.2 Microarray

DNA microarray offers the latest technological advancement for multiple-gene detection and pathogen detection[7]. Microarrays were initially used to examine gene expression for large numbers of genes, but because of it flexibility and high throughput capabilities microarray can be potentially used for DNA sequence analysis, immunology, genotyping and diagnostics. For pathogen detection on array, targets used can be PCR products or genomic DNA, into which reporter molecules (e.g. biotin and Cy-3) is incorporated directly via 5' primer modifications, nick translation, random priming, or chemical incorporation. Many combinations of fluorescent dyes and conjugates are suitable for detecting hybridized targets on microarrays. The labeled targets are then allowed to hybridize with the probes on the array plate. The hybridization signals are then measured by the various imaging techniques. The main advantage of microarray-based detection is that it comprises of powerful nucleic acid amplification strategies with the massive screening capability, which results in high level of sensitivity, specificity, and throughput capacity.

11.2.3.3 Biosensors

Biosensor technology is the latest development in pathogen detection[8]. Biosensors can detect small amount of samples with few rare false positive results and significantly reduce detection time. Biosensors used in pathogen detection are based on DNA hybridization. The target gene sequence is identified by hybridization to a DNA probe. The biosensor is an analytical device which incorporates biologically derived sensing element which is either integrated or intimately associated with a physicochemical transducer. Due to the specificity of the biosensor it may be used in complex media such as blood, serum, urine, fermentation broths, and food, often with minimum sample treatment. Both optical and electrochemical biosensors can be used for pathogen detection. Variations of biosensors such as amperometric biosensors, potentiometric biosensors, peizoelectric biosensors have been developed for faster and more accurate pathogen detection.

11.3 Pathogen detection in AlleleID

AlleleID facilitates the automatic design of primers and probes. This includes taxa specific probes, species specific probes and cross species probes. AlleleID analyzes the results of a multiple sequence alignment obtained from ClustalW and then design primers and probes.

11.3.1 Multiple Sequences Alignment

The purpose of multiple sequence alignments is to identify regions of similarity and difference between multiple related sequences. An optimal alignment of sequences is achieved when the greatest numbers of similar bases are aligned in the same column. Determining the optimal alignment of more than two sequences includes matches, mismatches and gaps is algorithmically intensive. Once the sequences are aligned, it is possible to determine a majority and minority consensus.

- **Majority consensus:** The majority consensus consists of the set of bases which occur in the largest number of sequences in a given position. This is used frequently in phylogenetic studies.

- **Minority consensus:** The minority concensus is the degenerate consensus in which only bases which appear in every member sequence will appear in the consensus. Otherwise a degenerate bases representing all possible bases will appear. This allows for easy identification of conserved regions for the design of cross species probes or taxa specific probes and point mutations for the design of species specific probes.

11.3.2 ClustalW

Many algorithms exist for aligning DNA and protein sequences but the most popular is the ClustalW algorithm[9]. Minimally, two sequences are required as input. All non-alphabetic characters (spaces, digits, punctuation marks) are ignored, except "-" which is used to indicate a GAP ("." in MSF-RSF). The program uses the input parameters, aligns the sequences, and displays the alignment. The alignment can then be used to generate a phylogenetic tree. The parameters used for multiple sequence alignment are:

- **Gap Penalties:** Controls the cost of opening up every new gap and the cost of every item in a gap. Increasing the gap opening penalty will make gaps less frequent. Increasing the gap extension penalty will make gaps shorter. Terminal gaps are not penalized.

- **Delay Divergent Sequences:** Delays the alignment of the most distantly related sequences until after the most closely related sequences have been aligned. The setting shows the percent identity level required to delay the addition of a sequence. The sequences that is less identical than this level to any other sequences will be aligned later.

- **Transition Weight:** The transition weight gives transitions (A <—> G or C <—> T, i.e. purine-purine or pyrimidine-pyrimidine substitutions) a weight between 0 and 1. A weight of zero means that the transitions are scored as mismatches, while a weight of 1 gives the transitions the match score. For distantly related DNA sequences, the weight should be close to zero; for closely related sequences it can be useful to assign a higher score.

11.3.3 Input format

In AlleleID, both multiple and single sequences can be imported from GenBank and from a file. It also supports the input of aligned Clustal files. Multiple sequences imported can be aligned and an enhanced sequence alignment view is generated by the AlleleID.

11.3.4 Primer and probes designed by AlleleID

AlleleID analyzes the results of alignment and design the primers and probes accordingly. AlleleID designs primer and probe sets for several applications:

11.3.4.1 Taxa specific oligonucleotides

For taxa specific probe design regions which are conserved among the members of the taxa but not present in other taxa are used for the primer and probe design. Common primers and probes[10] sets are designed to selectively amplify the DNA sequences of organism belonging to same taxa in a mixture. These common probe and primer avoid the need of designing different oligonucleotides for each and every individual sequence.

AlleleID supports the primer and probe design for both intergroup or taxa and intragroup. There are four different taxa of nematodes, Cephalobidae, Panagrolaimidae, Plectidae and Rhabditidae. AlleleID can design primers which are specific to Cephalobidae and Panagrolaimidae and do not amplify the DNA of organism belonging to Plectidae and Rhabditidae, thus supporting intertaxa primer design. It also supports the primer and probe design for discriminating between strains of similar taxa.

For Example: The **Panagrolaimidae** is subdivided into Panagrolaimus1 and Panagrolaimus2. The strains K1Panagrolaimus (A1), K107 UB8 1 EUC3 (F8) belong to Panagrolaimus1 and K171 UA5 2 EUC2 (B12) belongs to Panagrolaimus2. Primers designed [Table 11-1] will amplify K1Panagrolaimus (A1), K107 UB8 1 EUC3 (F8) and will not amplify K171 UA5 2 EUC2 (B12) in a reaction mixture [Figure 11-3].

Sense Primers	TTTGACGGATAACGGGGTATTAGG	59.1 °C
Antisense Primer	CACTACCTCCTCGTACTGAGAGT	59.1 °C
Antisense TaqMan	TTCGCGCCTGCTGCCTTCCTTGG	68.8 °C

Table 11-1. Represents the sequence, position and melting temperature for primers and common probe set for K1Panagrolaimus (A1), K107 UB8 1 EUC3 (F8).

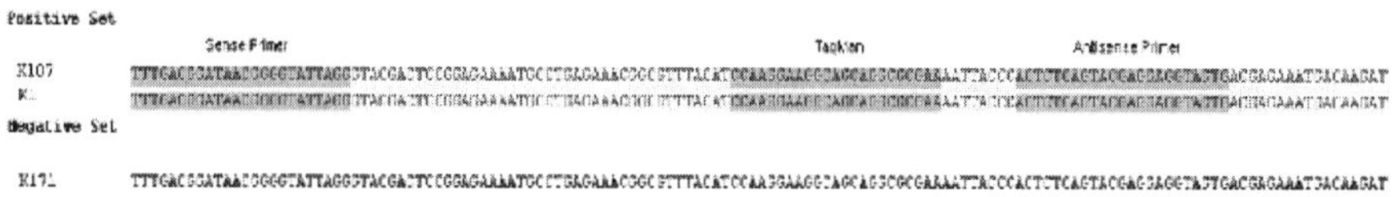

Positive Set

	Sense Primer	TaqMan	Antisense Primer
X107	TTTGACGGATAACGGGGTATTAGGGGTACGACTCCGGAGAAAATGCCTBAGAAACGGCBTTTACATCCAAGGAAGGCAGCAGGCGGGGGAAAATTCACCCACTCTCAGTAGCAGGAGGGTAFTGACGAGCAAATCACAAGAT		
K..	TTTGACGGATAACGGGGTATTAGGGGTTACGACTCCGGAGAAAATGCCTGAGAAACGGCTTTTACATCCAAGGAAGGCAGCAGGCGGGGGAAAATTCACCCACTCTCAGTAGCAGGAGGGTAFTGACGAGCAAATCACAAGAT		

Negative Set

X171	TTTGACGGATAACGGGGTATTAGGGGTACGACTCCGGAGAAAATGCCTBAGAAACGGCBTTTACATCCAAGGAAGGCAGCAGGCGGGGGAAAATTCACCCACTCTCAGTAGCAGGAGGGTAFTGACGAGCAAATCACAAGAT		

Figure 11-3.

Applications of taxa specific probes

a) **rRNA studies:** rRNA approaches are the major backbone of evolutionary studies. rRNA is generally used as it has a very slow rate of mutation and is conserved during the course of evolution. Studying rRNAs can give a lot of information about an organism's lineage and ancestral history. rRNA studies are also the basis of pathogen detection and biodiversity studies using PCR and hybridization probes.

b) **Biodiversity studies:** During the course of evolution, there are changes in the organism at the genomic level accounting for the diversity among different phylogenetic groups. In spite of these mutations in the genome, there are still certain regions of DNA which are conserved during evolution. Such conserved regions or sequences are useful in suggesting a phylogenetic group for an organism, identification of an organism or the extent of its biodiversity among the flora of a particular area. Taxa specific primer and probes are very useful in identification of a gene which ultimately leads to identification of new members of genomic families.

c) **Microbial Identification:** Classical microbial identification techniques rely on morphology, enzymatic conversion of substrates, or carbon source utilization, which provide subjective phenotypic identification. Microbial identification kits are based on the DNA sequence of ribosomal RNA genes, which is the new basis for bacterial taxonomic classification, leading to accurate and reproducible identification. Using 16S rRNA probes leads to accurate identification and classification of microorganisms. The methods which are generally in use are real time PCR and comparative PCR sequencing.

11.3.4.2 Species Specific Oligonucleotides

Phenotypic identification of organism cannot be done when the isolates of different organisms are being examined. These have to be identified genetically. Identification of species at the genomic level explains the variation in their structure and function. The alignment of DNA sequences is a useful tool in studying this genetic variation[11]. The basic information that the alignment provides is the identification of conserved and unique regions. This is useful in designing experiments to predict the function of specific genes and identifying new members of a gene family. Species identification primers are designed on the unique regions of a

247

sequence and hence can be used to amplify a specific species from a pool of templates. AlleleID designs two types of primers and probe sets for species identification unique probe and primer set, and minimal probe set.

11.3.4.2.1 Unique probe and primer set

Primers and probes designed are unique to a particular sequence in a group of sequences. To design these unique probe sets, all sequences are aligned and then regions unique to a species are identified.

For example, to identify three species of Geobacter; Geobacter grbicium(AF335183), Geobacter pelophilus (U96918) and Geobacter metallireducens(L07834). Three unique sets of primers and probes are designed [Figure 11-4 and Table 11-2].

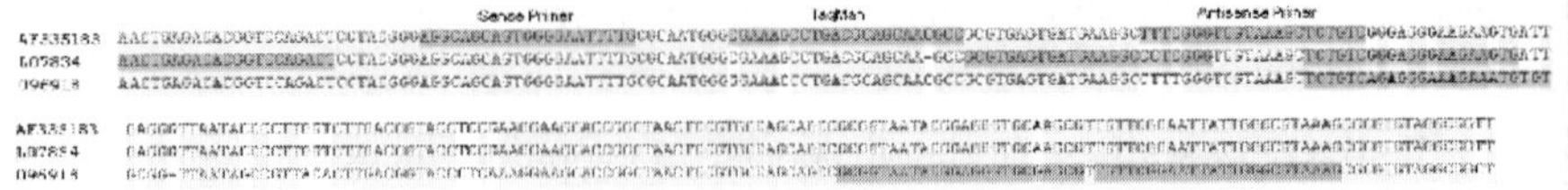

Figure 11-4.

Species	Position	Sequence	Length	Melting T_m
AF335183				
Sense	353	AGGCAGCAGTGGGGAATTTTG	21	59.7 °C
Antisense	444	GACAGAGCTTTACGACCCGAAA	22	59.1 °C
Sense TaqMan	383	CGAAAGCCTGACGCAGCAACGCC	23	68.1 °C
L07834				
Sense	324	AACTGAGACACGGTCCAGACT	21	59.0 °C
Antisense	459	CACTTCTTCCCTCCCGACAGA	21	59.4 °C
Antisense TaqMan	429	CCCGAGGGCCTTCATCACTCACGC	24	68.2 °C
U96918				
Sense	439	TCTGTCAGAGGGAAAGAAATGTGT	24	58.8 °C
Antisense	580	CTTTACGCCCAATAATTCCGAACA	24	58.9 °C
TaqMan	555	CGCTCGCACCCTCCGTATTACCGC	24	68.5 °C

Table 11-2. Represents sequence position and melting temperature for primers and unique probe sets for Geobacter grbicium(AF335183), Geobacter pelophilus (U96918) , and Geobacter metallireducens(L07834).

11.3.4.2.2 Minimal probe set

Rather than designing a primer and probe set for every target sequence, AlleleID provides an algorithm that will generate the minimum number of probes to uniquely identify every member of the alignment. This results in a cost savings in both probe synthesis and the number of reactions per assay. To do this, a matrix is generated showing which probe will bind to which DNA sequence.

For Example : For amplifying MHC class I heavy chain of all 5 chicken MHC cell lines minimum number of probes sets [Table 11-3] and matrix [Table 11-4] is generated by aligning the sequences together [Figure 11-5]

Target Sequence	Primer/Probe	Sequence	Position	T$_m$ (°C)
BFII15	Sense Primer	TCACAGCGCACTGAGCAGAT	262	60.2
	Antisense Primer	CGAGGATGTCACAGCCGTACA	373	60.4
	Sense TaqMan (P0)	ACCGCGATGGCCTGGGCACAC	284	68.4
BFII13	Sense Primer	GCTCCATACCCTGCGGTACA	69	59.7
	Antisense Primer	GAAGAGTTCCCCGTCCACATAC	165	58.9
	Antisense TaqMan (P1)	ACAGTCACGAACCACGGCTGCCC	140	68.2
BFII12	Sense Primer	GACGCCATGCAGGGAAGAAG	974	59.4
	Antisense Primer	CAAATGCTGGTGTGGACTGTTG	1119	59.1
	Sense TaqMan (P2)	ACAACATCGCGCCCGACAGGGA	1004	67.4
BFII5	Sense Primer	ACCCTGCGGTACATCCAAAC	76	58.3
	Antisense Primer	AAGAGTTCCCCGTCCACGTA	164	58.5
	Antisense TaqMan (P3)	ACAGTCACGAACCACGGCTGCCC	140	68.2
BFII2	Sense Primer	GACGCCATGCAGGGAAGAAG	974	59.4
	Antisense Primer	CAAATGCTGGTGTGGACTGTTG	1119	59.1
	Sense TaqMan (P4)	ACAACATCGCGCCCGACAGGGA	1004	67.4

Table 11-3. Represents the probes and primers set for 5 MHC chicken cell lines BFII15, BFII13, BFII12, BFII5 and BFII2.

Sequences	P0	P1	P2	P3	P4	Status
BFII15	–	–	–	–	+	Yes
BFII13	–	+	–	–	–	Yes
BFII12	+	–	+	–	–	Yes
BFII5	–	–	–	+	–	Yes
BFII2	+	–	–	–	–	Yes

Table 11-4. Represents the minimal probe matrix for 5 MHC chicken cell lines BFII15, BFII13, BFII12, BFII5, BFII2.'+' sign indicates that probe will bind to sequence and '-'indicated that given probe will not bind to sequence.

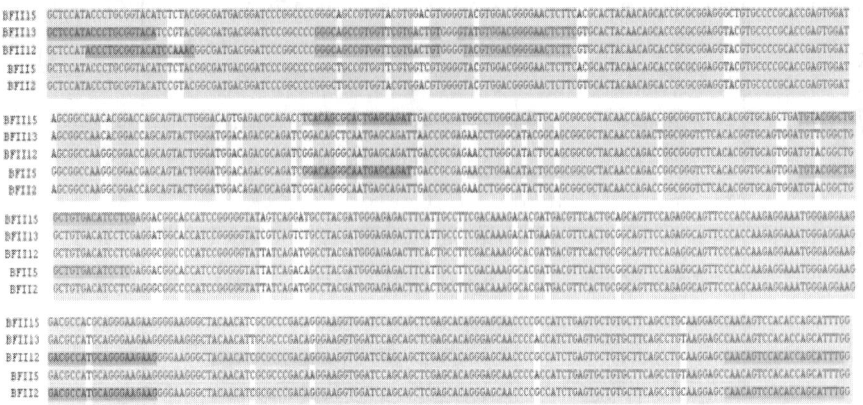

Figure 11-5.

11.3.4.3 Cross Species Oligonucleotides

During studies of a target gene, it is often desirable to study multiple organism models[12]. One method is to design and synthesize separate probes for each target organism. But by designing a single set in conserved regions of the gene, the same set can be used for all targets. In addition, the probe has a high probability of working with organisms whose sequence is unknown since the probe is located in a conserved region. Microarrays are one of the common tools used for the cross species analysis. Cross-species microarrays are finding great applications in studying genomes whose sequences are not known. The conserved regions identified is very useful in designing experiments to test and modify the function of specific genes, in predicting the function of genes and in identifying new members of genomic families. AlleleID uses the identified conserved region from a multiple sequence alignment and then designs both microarray and qPCR primers and probes that will work for all target sequences. [Figure 11-6] [Table 11-5]

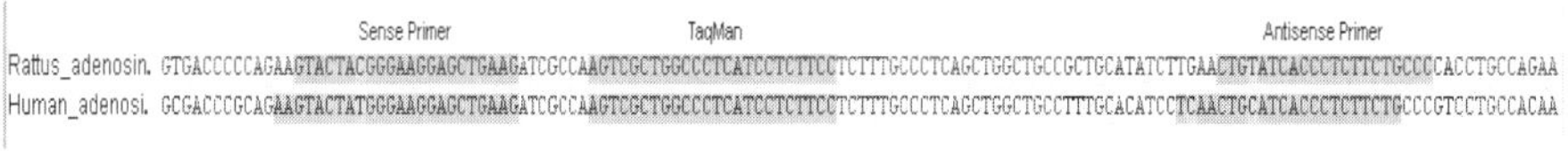

Sense Primer   TaqMan   Antisense Primer

Rattus_adenosin. GTGACCCCCAGAAGTACTACGGGAAGGAGCTGAAGATCGCCAAGTCGCTGGCCCTCATCCTCTTCCTCTTTGCCCTCAGCTGGCTGCCGCTGCATATCTTGAACTGTATCACCCTCTTCTGCCCCACCTGCCAGAA

Human_adenosi. GCGACCCGCAGAAGTACTATGGGAAGGAGCTGAAGATCGCCAAGTCGCTGGCCCTCATCCTCTTCCTCTTTGCCCTCAGCTGGCTGCCTTTGCACATCCTCAACTGCATCACCCTCTTCTGCCCGTCCTGCCACAA

Figure 11-6.

Accession Number	Primer/Probe	Sequence	Position	Tm (°C)
L22214	Sense Primer	AAGTACTATGGGAAGGAGCTGAAG	1109	58.5
	Antisense Primer	CAGAAGAGGGTGATGCAGTTGA	1218	58.8
	Sense TaqMan	AGTCGCTGGCCCTCATCCTCTTCC	1140	66.8
NM_017155	Sense Primer	GTACTACGGGAAGGAGCTGAAG	1111	58.4
	Antisense Primer	GGGCAGAAGAGGGTGATACAG	1221	58.1
	Sense TaqMan	AGTCGCTGGCCCTCATCCTCTTCC	1140	66.8

Table 11-5. Represents the cross species primer and probe sets for the L22214 and NM_017155.

For Example : Common primer and probes [Table 11-5] which identify adenosine A1 receptor (ADORA1) gene both in Human and Rats [Figure 11-6].

11.4 General Probe and Primer Design Considerations

AlleleID supports the design of TaqMan® probes and molecular beacons for quantitative detection of pathogens. The general guidelines used for primer and probe designs are:

11.4.1 Primer design guidelines

1. Primers should be 17-28 bases in length.
2. Primer base composition should be 50-60% (G+C).
3. Primers should end (3') in a G or C, or CG or GC: this prevents "breathing" of ends i.e prevents instability of the primers and increases efficiency of priming;
4. Tm should be between 55-80 °C.
5. 3'-ends of primers should not be complementary to reduce the formation of primer dimer competing for primer availability in the reaction
6. Primer self-complementarity (ability to form 2 °C structures such as hairpins) should be avoided.
7. Runs of three or more Cs or Gs at the 3'-ends of primers may promote

mispriming especially in GC-rich sequences and should be avoided.

8. There should be maximum of 2 °C difference in Tm between the two primers.

11.4.2 Probe Design Guidelines

11.4.2.1 Molecular beacon design guidelines

To successfully monitor PCR reactions, molecular beacons should be designed so that they are able to hybridize to their targets at PCR annealing temperatures. Simultaneously, the free molecular beacons should stay closed and be nonfluorescent at these temperatures.

1. The molecular beacon probe region should be 15 to 33 nucleotides long.
2. Its melting temperature should be 7-10° C higher than the PCR annealing temperature.
3. The loop sequence must be complementary to the target sequence of the assay.
4. It is important to design the molecular beacon in an area where there is minimal secondary structure formation of the target. This will help prevent the template from preferentially annealing to itself faster than to the molecular beacon.
5. The molecular beacon should bind at or near the center of the amplicon.
6. The distance between the 3'-end of the upstream primer and the 5'-end of the molecular beacon (stem) should be greater than 6 nucleotides.
7. The melting temperature of the probe-target hybrid can be predicted using the percent GC rule. The prediction should be made for the probe sequence alone before adding the arm sequences.
8. The stem sequences should not be complementary to the target sequence.
9. This stem region of the molecular beacon should be 5 to 7 bp long, with the GC content at 70-80%.
10. The length, sequence and GC content of the stem should be chosen such that the melting temperature is 7-10°C higher than the annealing temperature of the PCR primers.
11. A 5 bp stem will melt at 55-60°C, a 6 bp stem at 60-65° C, and a 7 bp stem at 65-70°C.
12. Since the G nucleotide may act as a quencher, it is best to avoid designing molecular beacon with a G directly adjacent to the fluorescent dye (typically at the 5' end of the stem sequence) and therefore cytosine at the 5' end of the stem is preferred.
13. The presence of any secondary structure other than the hairpin stem can change the position of the fluorophore with respect to the quencher causing background signals.
14. Using extremely long stems will make the molecular beacon inefficient in binding to its target.
15. The specificity of the probe can be relaxed by making the probe sequence in the loop longer than the usual 21 nucleotide average size. Such "mismatch-tolerant" molecular beacons offer greater flexibility in targets.

11.4.2.2 TaqMan® Probe design considerations

1. The probe melting temperature should be 10°C higher than the melting temperature of the primers.
2. 20-80% GC content
3. Length 9-40 bases (Preferably 20-30 bases long).
4. No G on the 5' end. A 'G' adjacent to the reporter dye quenches reporter fluorescence even after cleavage.
5. Fewer G's than C's.
6. The 3' end of the probe must be protected against chain elongation during PCR. For blocking the 3 'end phosphate, cordycepin, 2',3'-dideoxynucleosides, inverse T or the quencher dye itself can be used. Labeling the 3'end with TAMRA or another quencher can be achieved by using a 3'amino-linker.
7. Amplicon size 50-150 bp.
8. 3 ' end of primer as close to probe as possible without overlapping.

11.5 Conclusion

Pathogen identification has become a routine diagnostic performed in numerous clinical and research laboratories. Improvements in molecular technique have given them a distinct advantage over conventional techniques due to improvements in speed, accuracy, and the variety of pathogens that can be detected. The design of molecular systems manually is a tedious task and generally results in primers and probes that have not been optimized. AlleleID automates these tasks enabling the automatic design of taxa specific, species identification, and cross species primer and probes that follow standard design guidelines.

References

1. Feng, P.: Rapid Methods for Detecting Foodborne Pathogens. In: FDA Bacteriological Analytical Manual 2001, http://www.foodinfonet.com/working/publication/fdaBAM.htm.
2. Alvarez, A.M.: Integrated approaches for detection of plant pathogenic bacteria and diagnosis of bacterial diseases. *Ann. Rev. Phytopathol.* 2004 ; 42: 339-366.
3. Tyagi, S., Kramer, F.R. :Molecular Beacons : Probes that fluoresce upon Hybridization. *Nat. Biotechnol.* 1996;14:303-308.
4. Holland, P.M., Abramson RD, Watson R, Gelfand DH : Detection of specific polymerase chain reaction product by utilizing the 5'——3' exonuclease activity of Thermus aquaticus DNA polymerase. *Proc. Natl. Acad. Sci.* 1991; 88:7276-80.
5. Vet, J.A., Majithia AR, Marras SA, Tyagi S, Dube S, Poiesz BJ, Kramer FR:Multiplex detection of four pathogenic retroviruses using molecular beacons *Proc. Natl. Acad. Sci.* 1999; 96:6394-6399.
6. Leutenegger, C.M.:The Real-Time TaqMan® PCR and Applications in Veterinary Medicine. *Veterinary Sciences Tomorrow* 2001;1:1-15.
7. Calla, D.R., Boruckia, M.K., Loged, F.J.: Detection of bacterial pathogens in environmental samples using DNA microarrays. *J. Microbiol. Methods* 2003;53:235-43.
8. Tuan, Vo-Dinh: Biosensors, Nanosensors And Biochips Frontiers In Environmental And Medical Diagnostics Oak Ridge National Laboratory. 1st *International Symposium on Micro & Nano Technology* 2004.
9. Thompson, J.D., Higgins, D.G., Gibson, T.J.: CLUSTAL W: improving the sensitivity of progressive multiple sequence alignment through sequence weighting, position specific gap penalties and weight matrix choice. *Nucleic Acids Res.* 1994;22:4673-4680.
10. Walter, J., Hertel, C., Tannock, G.W., Lis, C.M., Munro, K., Hammes, W.P.: Detection of Lactobacillus, Pediococcus, Leuconostoc, and Weissella Species in Human Feces by using Group-Specific PCR Primers and Denaturing Gradient Gel Electrophoresis. *Appl. Environ. Microbiol.* 2001;67: 2578–2585..
11. Whitby, P.W., Pope, L.C., Carter, K.B., LiPuma, J.J., Stull, T.L. Species-Specific PCR as a Tool for the Identification of Burkholderia gladioli. *J. Clin. Microbiol.* 2000;38:282–285.
12. Frazer, K.A., Elnitski, L., Church, D.M., Dubchak, I., Hardison, R.C.: Cross-species sequence comparisons: a review of methods and available resources. *Genome Research* 2003;13:1-12.

Chapter 12

Oligonucleotide Design from Principle to Practice

Paul M.K. Gordon, Christoph W. Sensen*

Contact: Christoph W. Sensen, Ph. D.
University of Calgary, Sun Center of Excellence for Visual Genomics,
Faculty of Medicine, Department of Biochemistry and Molecular Biology,
3330 Hospital Drive NW, Calgary, Alberta, Canada, T2N 4N1
Email: csensen@ucalgary.ca

Abbreviations: AS – antisense, CDS – coding sequence, NN– nearest neighbours, oligo – oligonucleotide, ORF – Open Reading Frame, PCR – polymerase chain reaction, PSSM - Position Specific Scoring Matrix, siRNA – small interfering RNA

12.1 Introduction

The term "oligonucleotide" (oligo) is normally used to designate a laboratory-fabricated DNA or RNA strand for use in a Molecular Biology or Genomics experiment. Virtually all applications of oligonucleotides involve synthesizing the complementary strand of a targeted, naturally occurring, strand of nucleic acid sequence. In the experiment, the synthesized oligo will bind to the targeted sequence according to classic Watson-Crick base-pairing rules to form a double stranded nucleic acid molecule[1]. The oligo base sequence is therefore fully determined by the targeted sequence moiety of the duplex.

Because the length of an oligonucleotide is directly proportional to its synthesis cost, and inversely proportional to its efficacy (as explained later on), the desired oligo length is usually only a fraction of the targeted sequence's length. The experimenter therefore has many target subsequences to consider as candidates for the oligo binding region. Determining the most effective oligo candidate can involve many metrics that can be calculated several ways, therefore a sound knowledge of oligo design principles is essential in order to achieve a good experimental design.

Experiments that use oligos can be divided into two types depending on the fate of the target sequence: (a) When oligos are used to start a chain reaction on the target, the oligos are often referred to as "primers"; (b) when the oligo is used to bind and hold the target, the oligos are often referred to as "probes". Primer applications are those requiring a Polymerase Chain Reaction (PCR) product, such as DNA sequencing, and cDNA microarrays. Probe applications include Northern and Southern blots, antisense gene expression inhibition and oligonucleotide microarrays. This chapter outlines both the theoretical and practical aspects of optimal oligonucleotide design for various laboratory applications, assuming that the reader has a basic knowledge of principles behind PCR, DNA sequencing and nucleotide microarrays.

12.2 Oligo Suitability Metrics

Before examining methodology and quantification in detail, we should understand the motivation behind each oligo suitability metric. All lab applications described below share some common metrics for oligo suitability. In addition, restrictions on the oligo's effective biochemical characteristics and binding location depend on its intended use.

12.2.1 Metrics used for all applications

12.2.1.1 Hairpins & Dimers

The probe or primer function of the oligo is dependent on the availability of the nucleotide bases for pairing with the target. Two measures that are critical in oligo design are: (a) the propensity of the oligo to fold into a hairpin structure joining the two ends; and (b) the propensity of two copies of the oligo to bind to each other through self-complementarity to form a dimer. Both situations make the oligo's bases effectively unavailable for target binding, and should be avoided (with the rare exception of stem-loop oligo protocols[2]).

12.2.1.2 Target availability

Paralleling the effects of oligo hairpin and dimer structures, secondary structures in the target (e.g. mRNA or cDNA) can interfere with probe binding. For the same reason, the probe should not target known protein-DNA binding motifs. Subtle changes in base pairing can greatly affect binding efficiency. One should also avoid, if possible, targeting regions with multiple known single nucleotide polymorphisms (SNPs), unless the characterization of SNPs is the subject of the experiment.

12.2.1.3 Melting Temperature

For primer applications, selecting an oligo with an appropriate melting temperature is key to ensuring that the primer denaturing phase of PCR can take place. The denaturing of the two strands of the duplex is not a neat two-phase transition, but rather consists of multiple intermediate duplex configurations. The melting temperature, T_m, is generally defined as the point at which half of the duplexes are annealed. Annealing temperature for PCR is generally considered to be 5°C below melting temperature. PCR products are most successfully produced when the 5' and 3' PCR primers have similar melting temperatures. A maximum difference of 2°C is generally attainable in the design process by adjusting the length of one primer.

12.2.1.4 Oligo Length

With the exception of siRNAs (see section 12.2.2.2), the length of the oligo itself is not extremely critical to the success of oligo experiments. The constraints on oligo length are normally dictated by the G+C composition of the target, and the desired temperature range. Oligo length is inversely proportional to G+C, and directly proportional to temperature. For PCR primers, generally 18-30mers are used. For oligonucleotide probe microarrays, generally 60-80mers are used. An additional consideration for microarray oligonucleotide length constraints is that commercial oligo synthesizers, due to manufacturing process limitations, may restrict the length of oligos available for any given quantity ordered. For example, when ordering 50 nmol of an oligo, the synthesis may be limited to oligos of a maximum length of 60 nucleotides.

12.2.1.5 Base distribution

Most oligo design guides will suggest avoiding homogeneous (4 or more base) GC regions in the oligo[3]. The stable triple-bonded G:C pairs attenuate the multistage melting transition by their reluctance to separate even after the rest of the oligo duplex has denatured.

Avoiding probes with many consecutive adenosines (As) is also desirable for oligo microarrays. The cDNA sample is typically labelled via attachment to modified uracils (Us), and a quenching effect occurs when typical labels such as Cy5 and Cy3 are placed closer than 6 bases apart.

12.2.1.6 Off-target Binding

In most applications, repetitive genomic sequences (repeats) are undesirable in oligo duplexes because the oligo is intended to bind to a unique target. Repeat regions should therefore be masked out of the candidate binding site list. Conversely, RNA arbitrarily primed PCR (RAP-PCR) is one case in which repeats are desirable: the objective is to form as many different PCR products from a cDNA population as possible, for differential display. In cases such as this, the frequency of substrings in the known coding parts of the genome must be determined.

Oligos may also bind to non-exact complementary sequences with a lesser but still significant degree of stability. This can cause nonsense reads in the case of sequencing primers, multiple products in the case of PCR amplification, and false gene expression levels readings in the case of oligo microarrays. The acceptable amount of off-target binding stability (also called secondary binding) should normally be relative to the perfect complement duplex stability, as discussed in the thermodynamics section of this chapter.

12.2.2 Application Specific Metrics

12.2.2.1 Positional Constraints

The position of oligo duplexes on the target sequence can affect the quality of results for PCR products, sequencing, and hybridization. A common consideration for eukaryotic PCR products and DNA hybridization experiments is a 3' template bias, because 3' cDNA is over-represented when polyadenylated eukaryotic mRNA template is reverse transcribed using poly(T) primer. For prokaryotes, typically a random hexamer mix is used to prime reverse transcription; there exists a slight bias towards the 5' end of the cDNA, since reverse transcription can start anywhere and then proceeds towards the mRNA template's 5' end. Also, the distance between a PCR primer pair's 5' and 3' binding sites on the template DNA should normally be less than 1500 bases, to ensure that the product size is not larger than the standard limits of viable product extension. Be aware that the extension limit varies directly with the duration of the primer extension phase in the PCR protocol used. Some of the many specialized PCR protocols also have additional position constraints, such as nested PCR where additional oligos must fall inside the primary PCR product sequence. PCR protocols where primers must span exon boundaries provide the least amount of design flexibility.

The position range of directed sequencing primers is dependent on the quality of the known template primed, and the performance of the sequencing machine used. Sequencing primers are divided into two categories: (a) "walking" primers that elucidate bases off the end of a known template; and (b) "polishing primers" that elucidate ambiguous parts of a known template (either due to poor quality or polymorphism). The candidate range for primers is determined by the following formulae:

Primer range end = T - U - O - L

Primer range start = range end - R + M

T = Target base, location on DNA strand that needs elucidation

U = Unreadable bases after oligonucleotide binding site (due to sequencing chemistry, typically 10-20)

O = required assembly Overlap bases (walking primers only)

L = oligonucleotide Length, specified by the user

R = sequencing Read length, minimum expected

M = Minimum number of post-T bases to be elucidated

This formula is depicted graphically in Figure 12-1. Without the M parameter, the best walking primer may result in only 5 bases beyond the already existing sequence! If no candidate meets the M requirement, the user can change parameters such as secondary binding to be less stringent; or simply extend other contigs until a walk from another contig joins the troublesome end.

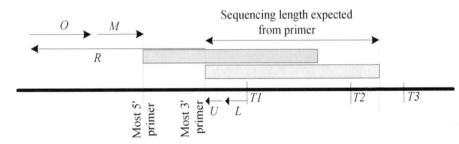

Figure 12-1. *Range of candidate sequencing primers for elucidation including position T1 (e.g. an assembly ambiguity to resolve), with user specified parameters. Candidates are checked from 3' to 5' within the range to minimize the number of primers required. An early candidate could also resolve T2 in the same reaction. U = number of unreadable bases. L = primer length. R is expected sequence read length. M is minimum number of bases to elucidate after T1. Without M, a walking primer could elucidate only one new 3' base.*

12.2.2.2 siRNA Constraints

RNA interference (RNAi) is a natural eukaryotic mechanism (possibly a primitive immune system against dsRNA viruses) for mRNA depletion, and therefore suppression of gene expression. The small interfering RNA (siRNA) mechanism involves an RNA-induced silencing complex (RISC) that searches for 21-23mer double stranded RNA (dsRNA) that has been processed by the Dicer endonuclease.

The dsRNA is unwound, and the antisense strand (AS) is used to guide the RISC to complementary mRNA, which is then cleaved[4]. Formulating a dsRNA for siRNA activity requires two complementary RNA oligonucleotides with dinucleotide overhangs. The actual overhang sequences appear to have little effect, only their presence is necessary. The AS sequence should be completely complementary to the targeted mRNA to maximize this targeted seek-and-destroy activity, but additional rules are evolving to improve the efficiency of the RNAi effect.

Sequence patterns in the siRNA seem to affect the probability of RISC unwinding the dsRNA and incorporating the AS to form an active complex. In effective siRNAs, RISC preferentially unwinds the dsRNA from the 5' AS end. To promote this activity, the AS RNA oligo should be designed with relative unstable (A/U) pairings at its 5' end, and strong (G/C) pairings at its 3' end. There is also evidence that relative instability at the cleavage site (between bases 10 and 11) improves siRNA efficiency [5]. These tendencies are confirmed by patterns observed in microRNAs (miRNA), siRNAs' endogenous counterparts. An additional factor to be considered when designing siRNAs is peculiar off-target gene silencing. There is some evidence that mRNAs with as few as 11 contiguous matches to the AS are significantly silenced, and that non-canonical G:U base pairing between the AS and target is tolerated. Currently, some published methods[6, 7] conflict in their siRNA rules, evidence that this is an evolving field.

12.3 DNA Thermodynamics

We have already referred to duplex stability and melting temperature in the suitability metrics above, but how are these determined for nucleic acids? The classic physical chemistry formulae for thermodynamic energy have been adapted to deal with nucleic acid base pairing. Knowing how these energies are measured will give us a better idea of how to set appropriate thresholds for the suitability metrics.

12.3.1 Rules of Thumb

Many rules of thumb exist for estimating oligo duplex T_m, the most famous being the so-called Wallace rule[8] for short oligos in 0.05M monovalent cations:

$$T_m = 4 \cdot (\# \; C+G \; bases) + 2 \cdot (\#A+T \; bases)$$

Other empirically derived formulae take into account Na+ and K+ concentrations for non-PCR applications[9] (PCR is typically performed with 0.05M monovalent cations), such as:

$$T_m = 81.5°C + 16.6°C \cdot \log_{10}([Na+] + [K+]) + 0.41°C \cdot (\%G+C) - 675/(oligo \; length)$$

These methods are useful only when the oligo application does not require a precise melting or annealing temperature determination. If more precise estimates of melting temperature are required, a thermodynamic model must be used. Such models involve more complex calculations and value lookup tables, and therefore computer programs must be used rather than rules of thumb.

12.3.2 Thermodynamics and the Nearest Neighbour (NN) Model

The NN model is generally accepted as the most precise way to predict the thermodynamic properties of DNA and RNA duplexes. It is based on the long-standing premise that most of the binding energy in a duplex is derived from the interaction of adjacent base pairs[10]. The amount of enthalpy (order, ΔH^o) and entropy (disorder, ΔS^o) intrinsic to these neighbour interactions can be summed to provide overall enthalpy and entropy values for the whole duplex. The melting temperature in degrees Kelvin can then be calculated as:

$$T_m = \Delta H^o/[\Delta S^o + R \cdot \ln(C_T/4)]$$

Where R is the molar gas constant, and C_T is the concentration of the oligo. From the same enthalpy and entropy values, the stability of a duplex can be measured in terms of free energy (ΔG) at a given temperature T (in degrees Kelvin):

$$\Delta G^o{}_T = \Delta H^o - T \cdot \Delta S^o$$

Negative free energy indicates a stable structure. Increased salt ion concentration increases entropy, while enthalpy remains constant. Because ΔS^o is normally a negative value for DNA duplexes, the net consequence for these two formulae is that salt increases both melting temperature and stability. SantaLucia's NN model data [11] is based on a 1M NaCl environment, with the salt correction being:

$$\Delta S^o(\text{oligo, [Na+]}) = \Delta S^o(\text{oligo, 1M NaCl}) + 0.368 \cdot (\text{oligo length-1}) \cdot \ln[\text{Na+}]$$

The actual calculation of free energy and melting temperature for a perfect oligo duplex involves the formulae above using summed NN intrinsic enthalpy and entropy values. For 2.0mM of the oligo CCGAAA in 0.10 M NaCl at 37°C the NN derived data is:

$\Delta H^\circ(CCGTAAA) = \Delta H^\circ(CC)+\Delta H^\circ(CG)+\Delta H^\circ(GT)+\Delta H^\circ(TA)+\Delta H^\circ(AA)+\Delta H^\circ(AA)$
$= ((-8.0)+(-10.6)+(-8.4)+(-7.2)+(-7.9)+(-7.9))$ kcal/mol
$= -50$ kcal/mol
$\Delta S^\circ(CCGTAAA) = \Delta S^\circ(CC)+\Delta S^\circ(CG)+\Delta S^\circ(GT)+\Delta S^\circ(TA)+\Delta S^\circ(AA)+\Delta S^\circ(AA)$
$= ((-19.9)+(-27.2)+(-22.4)+(-21.3)+(-22.2)+(-22.2))$ cal/mol/K
$= -135.3$ cal/mol/K

$\Delta S^\circ(CCGTAAA, 0.1M\ NaCl) = -135.3 + 0.368{\cdot}6{\cdot}\ln(0.10)$
$= -140.4$ cal/mol/K

Note that entropy is usually measured in cal, not kcal, so either the enthalpy or entropy must be refactored for further calculation. The DNA concentration's natural logarithm is considered unitless. Also, NN thermodynamics are normalized to 37°C (310K). From these values we can calculate:

$T_m = -50000$ cal/mol / $[-140.4$ cal/mol/K $+ 1.987$ cal/mol/K${\cdot}\ln(0.00020/4)]$
$= 313K\ (40°C)$

$\Delta G^\circ = -50000$ cal/mol $- 310K{\cdot}\ (-140.4$ cal/mol/K$)$
$= 6.5$ kcal/mol

12.3.3 NN Model Data

SantaLucia's unified model (use in the worked example above), derived from the linear regression of 7 previous models, is generally considered the most accurate NN data for DNA. Since its publication in 1998, subsequent papers by SantaLucia and collaborators have added NN data for single base mismatches and dangling ends. Because oligos are almost always shorter than the targeted sequence, the 5' and 3' ends of the oligo often interact with their non-duplexed neighbours. Because the interactions can add as much stability as an A:T paring, inclusion of end interactions can be especially significant for short oligos (10-20-mers).

Data for single DNA base duplex bulges is also available for A and T neighbour contexts. While bulge and mismatch data are not essential for perfect duplex energy calculations, they are essential for proper prediction of off-target matches and oligo secondary structures. For RNA NN modeling, most data can be derived from the original Freier model, with subsequent updates from the Turner group.

12.3.4 Everything Is Relative

The actual values for melting temperature, secondary binding, and secondary structures are greatly affected by the G+C composition of the oligo. Absolute thresholds therefore do not make sense, but rather thresholds should be set relative to the characteristics of the perfect duplex. In *Sulfolobus solfataricus*, an archaeon with 35% G+C content overall in the genome, -13 kcal/mol (or less) of hairpin or dimer free energy for random 70mers is infrequent. In the human pathogen *Burkholderia pseudomallei* (70% G+C), virtually all optimal 70mers have less than -13 kcal/mol of hairpin or dimer free energy. A threshold rule of thumb is given in section 12.5.2. Melting temperatures can also be calculated for these secondary structures, which may be more intuitive for the end-user than free energy. A caveat is that these temperatures are a rather imprecise estimate for all but the simplest of structures, because of the complex ways in which folded nucleic acid complexes denature.

Given that the melting curve for DNA duplexes typically spans $T_m \pm 5°C$ under standard PCR conditions, a margin of at least $10°C$ between target duplex T_m and best off-target T_m is desirable to ensure no overlap in their curves.

12.3.5 Considerations for Microarrays

All of the thermodynamics models proposed for nucleic acids derive their results from oligonucleotide duplexes in an aqueous solution. Directly spotted microarrays typically anchor the oligo probes to a substrate via a linker molecule. The oligos will therefore anneal to the target in an environment with physical constraints and possibly non-uniform ion distribution due to substrate effects. These effects are largely ignored, or considered to be linear, therefore the aqueous measures are generally used as a guide, rather than an actual prediction of probe properties.

Another feature of oligonucleotide microarrays is that it is one of a few applications where a large number of probes will be annealing in the same environment. Under these circumstances it is beneficial to have similar thermodynamic properties for all of the probe duplexes, therefore the accurate prediction of melting temperature becomes especially important.

12.4 Computational Techniques

Understanding thermodynamics has given us insight into what units are used to measure oligo stability, and what factors affect it. Armed with this and the knowledge of what metrics are important, we are in a position to properly judge the techniques and threshold values that should be applied to oligo suitability.

12.4.1 Melting Temperature

There are literally hundreds of Web sites that provide melting temperature predictions for DNA and RNA oligos, employing a variety of methods. Table 12-1 lists some melting temperature and oligo services. SantaLucia's excellent HyTher Web server (http://ozone2.chem.wayne.edu/) calculates oligo duplex melting temperatures using the unified NN model and subsequent updates. Many Web sites and available command line programs (such as EMBOSS's eprimer3)[12] are wrappers around the Whitehead Institute's Primer3 software[13]. Primer3 allows the user to specify min/max thresholds for some of the criteria discussed so far. In addition, it has many other controls, making it a powerful tool in the hands of an experienced and knowledgeable user. The user should be aware that this program uses the (older and less accurate) Breslauer NN model[14]. In addition, thorough checks for hairpins, dimers, and secondary binding should be performed on the candidates produced.

Program Name	Web Site	Notable
HyTher	http://ozone2.chem.wayne.edu/	Reports melting temperatures, and best competitive hybridization
Primer3	http://www.genome.wi.mit.edu/cgi-bin/primer/primer3_www.cgi	Most commonly used program, father of many other oligo Web services
Osprey	http://osprey.ucalgary.ca/	Automates most parameter selection, sensitive off-target searches
CODEHOP	http://blocks.fhcrc.org/codehop.html	Generates primers for degenerative (multiple, related) sequence
Sfold	http://sfold.wadsworth.org/index.pl	Generates siRNA and antisense oligos
POLAND	http://www.biophys.uni-duesseldorf.de/local/POLAND/poland.html	Generate graphs of many duplex characteristics, can use different NN models

Table 12-1. *A selective list of oligo properties/design programs available on the Web.*

12.4.2 Secondary Structures

The most trusted tool by far for the calculation of secondary structures is the Mfold suite[15]. This suite includes a range of programs that predict stable DNA and RNA structures such as hairpins and dimers using the SantaLucia and Turner group free energy data. It can also be used to determine the melting temperature for a given duplex. Mfold is available over the Web from several bioinformatics service sites, or as source code from its author.

12.4.3 Off-target binding

Minimizing off-target binding consists of two steps: (a) repeat filtering; and (b) partial match detection. Several tools are available to filter (a.k.a. mask out) repetitive sequences, generally of 20 bases or greater. Where a lot of sequence for an organism is available, repeated substrings can be computed using RepeatMasker/MaskerAid[16] or RepeatFinder[17]. MEGABLAST[18] with query and database being identical can also identify repeats. Known repetitive sequences found in RepBase can be identified using BLAST. Most official genome project Web sites will also have a species-specific list of annotated repetitive sequences.

Repeat filtering should not be confused with low-complexity region filtering. Low complexity indicates an abnormally high likelihood of matches, and therefore can be useful if the complete sequence for the organism in question is not known. If the complete sequence is known, low complexity filtering may filter out regions that are in fact not significantly repeated. The NCBI's DUST is a popular low complexity filter.

Proper minimization of partial matches is a complex issue because off-target binding is governed by DNA thermodynamics, not simply complementarity. Many tools simply use a percentage identity cutoff from a BLAST search for off-target minimization. This has two problems: (a) first, percentage identity and thermodynamic energy are only moderately correlated; (b) second, BLAST cannot find all short, partial DNA matches, due to its implementation details. The first problem can be overcome by setting the maximum percentage identity threshold low, at the risk of excluding valid oligo candidates. The second problem can be partially overcome by lowering the "word size" and "mismatch penalty" parameters of BLAST, which slows BLAST down considerably in order to examine smaller, non-identical substrings than normally examined for matches. Finding the optimal match between two DNA or protein strings can be achieved using an implementation of the Smith-Waterman algorithm[19]. This is unfortunately a very computation-intensive task. Some Web sites, usually supported by a specialized bioinformatics hardware accelerator, can provide Smith-Waterman searches for the off-target datasets you require. Where Smith-Waterman is not available, FastA's pairwise alignment program[20] is preferable over BLAST's for short DNA searches. The FastA algorithm is a hybrid between the speed shortcuts of BLAST with the thoroughness of a Smith-Waterman. A novel alternative solution to both the thermodynamics and small match problems is discussed in the next section.

12.5 Putting It All Together: the Osprey System

Oligo design software that automates the criteria satisfaction process is chaining together calculation tools on the user's behalf. In this section, we will investigate the techniques behind Osprey, a software tool that spans the whole range of oligo design tasks. Osprey also contributes a novel means of dealing with off-target binding minimization. In this section the details of Osprey will be revealed to show how to find and use the "sweet spot" for parameters, where optimal oligonucleotides are more likely to occur. A Web interface to many of Osprey's functions can be found at http://osprey.ucalgary.ca/, and a complete discussion of its methodology and performance can be found in the 2004 publication[21].

12.5.1 The Process of Elimination

In Osprey, a set of all possible oligonucleotides is first created based on the user-selected target sequence range (for sequence assembly, amplification using genomic DNA as a target) and oligonucleotide size. Figure 12-2 illustrates how oligonucleotide candidates pass through a series of fitness exclusion tests, from fastest to slowest for efficiency. Osprey is designed to parallelize the processes wherever feasible to facilitate large-scale oligonucleotide selection. A configuration file containing biophysical parameters is used in Osprey, in addition to the command line specification of the files containing the sequence, assembly information and gene location. The default parameter values are meant to be reasonable approximations of the annealing conditions, but can and should be overridden using the command line or Web interface to match the conditions of individual experiments.

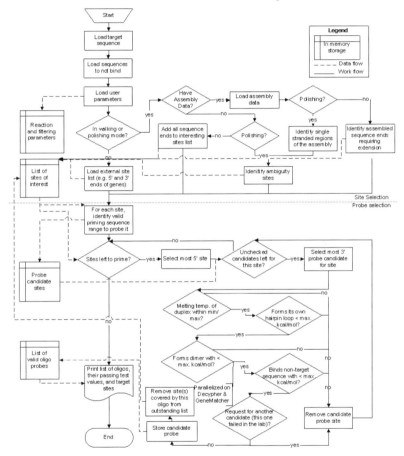

Figure 12-2. Workflow diagram of oligonucleotide primer selection in Osprey. Data flow links between "Reaction and Filtering parameters" and the test cascade in the lower right section are omitted for readability of the diagram. Sequences, assembly information and default parameters are read from disk, and many parameters and modes can be overridden on the command line. Note that the Probe Selection part of the processing is common to all oligonucleotide design modes. Also, the secondary binding check can be parallelized in Osprey using hardware accelerators.

267

12.5.2 Determining and Using Optimal and Threshold Criteria

Osprey's main program provides a mechanism to start the oligonucleotide design with tight constraints, and slowly loosens them if no appropriate oligonucleotides are found. The two parameters that can be adjusted automatically are melting temperature and oligonucleotide length. As an example, a spotted oligonucleotide microarray design may start with a temperature range of 78±5°C, and an oligonucleotide length range of 70±5 bases. All possible oligonucleotides with 70 bases and melting temperature of exactly 78°C will be found, and checked against the selection criteria. Targets that do not have any candidates passing the tests will be checked for 70mers with melting temperatures of exactly 77°C or 79°C. This process continues until all 70mers with 73°C or 83°C are checked, which is followed by the same checks for 71mers and 69mers, and so forth.

Selecting threshold values is a trade-off between having many candidates and having ideal candidates. One approach is to start with very strict criteria, find successful candidates, and if there are none, continue to loosen the constraints a bit at a time. For large-scale design such as a microarray or genome sequencing, such an approach can be very computationally expensive due to large numbers of targets analyzed over large numbers of iterations. The approach used by Osprey is three pronged: (a) first, formulae are used to estimate ideal start parameters for length and temperature based on G+C content and the lab application; (b) second, reasonable threshold values are defined for other criteria (e.g. hairpin energy) using rules of thumb — oligos satisfying all thresholds will be printed and removed from the list of targets to investigate further; (c) third, absolute thresholds for criteria are defined so that candidates exceeding the reasonable threshold but lower than the absolute threshold can be considered as "suboptimal" candidates.

This three-pronged approach mimics the mental process applied by an experienced human investigator to oligo selection, but it applies the process more systematically. The example of a 70mer oligo microarray for a 65% G+C organism illustrates the technique. The input is the set of transcripts for the organism, which Osprey filters for repeats using a wrapper program around MEGABLAST. Given that we want oligos of about 70 nucleotides in length, we can estimate the average 70-mer melting temperature by analyzing base distribution to determine average NN enthalpy and entropy (which we will denote by ΔH° and ΔS°, see Table 12-2),

DNA G+C%	Average NN enthalpy ($\Delta H°$)	Average NN entropy, 1M Na+ ($\Delta S°$)
35	-8050	21.93
40	-8120	21.99
45	-8190	22.06
50	-8280	22.13
55	-8350	22.19
60	-8430	22.26
65	-8510	22.33
70	-8590	22.40

Table 12-2. Average entropy and enthalpy NN estimates for DNA with various GC compositions.

and plugging values into the T_m formula previously described, with 0.1M NaCl and 0.0002M DNA:

$$T_m = \Delta H°\cdot N/(((\Delta S° + 0.368\cdot\ln(Na+))\cdot N) + R\cdot\ln(CT/4))$$
$$= -8510\cdot69/(((-22.3 + 0.368\cdot\ln(0.1))\cdot69) + 1.987\cdot\ln(0.0002/4))$$
$$= 363K \ (90°C)$$

where N is the length of the oligo minus one (i.e. the number of nearest neighbours). The opposite situation can also occur. Given a target temperature, say 90°C (363K), an average oligo length producing it can be determined by rearranging the above formula to solve for N+1:

$$N+1 = -R\cdot\ln(CT/4)/(\Delta S° + 0.368\cdot\ln(Na+) - \Delta H°/\ T_m) + 1$$
$$= -1.987\cdot\ln(0.0002/4)/(-22.3 + 0.368\cdot\ln(0.1) - (-8510/363) + 1$$
$$= 70$$

However, it should be noted that the latter formula is much less stable than the former in practice, and offers only a rough guide to the ideal length. Hairpin and dimer threshold are then set according to a rule of thumb that has proven effective over a range of organisms as not being either too inclusive or exclusive. The rough rule of thumb, where N is the oligo length, is:

$$10kcal/mol + integer((\%GC-35)/5) + N/20$$

Criteria thresholds indicate that a candidate satisfying them has no problems. Candidates investigated later, with different length or melting temperatures, may have slightly better overall criteria scores. The difference though would not have an impact on the experiment; therefore we stop the oligo search for this target with the discovery of this first satisfactory candidate. On the opposite end of the spectrum, some targets may not have any satisfactory candidates at all because they have, for example, an unusually high G+C composition compared to the overall target dataset.

Instead of throwing away all candidates above the satisfactory thresholds, a measure is calculated that says how far off a candidate is from being acceptable. The best (lowest scoring) suboptimal oligo will be printed when all valid oligo length and melting temperature range values have been exhausted. This avoids having to iteratively recalculate candidates for tough targets by manually tweaking thresholds. The formula used in Osprey to estimate closeness to acceptability is:

$$(\Delta \text{off-target melt})^2 + (\Delta \text{hairpin})^2 + (\Delta \text{dimer})^2 + (\text{GC clamp score})$$

Where each delta score is 0 when the criteria is acceptable, otherwise it is the rounded-off integer difference between the actual value and acceptable threshold. The GC clamp score is between 0 and 1 (with 1 being 10 consecutive G's and C's), therefore it only affects candidates that are otherwise equal. Given two candidates with 0.5 GC clamp score, suppose one exceeds the off-target melting threshold by 3.7°C, and the other exceeds the off target by 1.2°C, the hairpin by 0.6kcal/mol and the dimer by 2.2kcal/mol:

Candidate 1: $4^2 + 0^2 + 0^2 + 0.5 = 16.5$
Candidate 2: $1^2 + 1^2 + 2^2 + 0.5 = 6.5$

Candidate 2 is clearly closer to optimal by this formula. The squaring of terms in the formula favors a candidate that exceeds each threshold a bit rather than a candidate that exceeds any one threshold excessively.

12.5.3 Sensitive Off-Target Identification and Free Energy: A Novel Method

We have previously mentioned that the Smith-Waterman algorithm can be used to ensure that all small, partial matches can be found between the oligo and off-target sequences. While this solves the problem of sensitivity, there remains the problem of precision (how many of the hits found are actually real problems). A pairwise alignment is imprecise because of the merely moderate correlation between percentage identity and melting temperature. A more precise method would take into account the nearest neighbour phenomenon that underlies the biochemical model. A single pairwise match or mismatch, when combined with information about the base immediate 5', can have one of 4^3, or 64 energy states. In BLAST and Smith-Waterman searches, an adenosine matching an adenosine gets the same score, regardless of the neighbouring bases or the position of the match in the sequence. Such scoring is unsuitable for encoding the many thermodynamic neighbour states required for accurate computation. Osprey uses another sensitive search method, called a Gribskov profile[22], to find partially matching substrings, with a scoring mechanism based on neighbour relationships rather than pairwise identity.

Gribskov profiles are a form of position specific scoring matrices (PSSMs) used in bioinformatics. Profiles searches are compute-intensive, like Smith-Waterman searches, therefore several bioinformatics hardware accelerators (most notably TimeLogic's Decypher), are programmed to quickly perform PSSM searches. In PSSMs, each position in the query sequence has an independent score, for example two adjacent A's may be rewarded differently when matched. Free energy is measured in the Osprey profiles rather than using melting temperature. According to the formulae previously discussed, melting temperature calculation would require tracking two sets of scores, enthalpy and entropy. Profiles can only track one.

Figure 12-3 plots free energy vs. melting temperature, and indicates a strong correlation between melting temperature and ΔG, making ΔG a suitable candidate for selection and filtering of oligo specificity under experimental hybridization conditions. This makes sense because both the free energy and melting temperature formulae are based on the same enthalpy and entropy values.

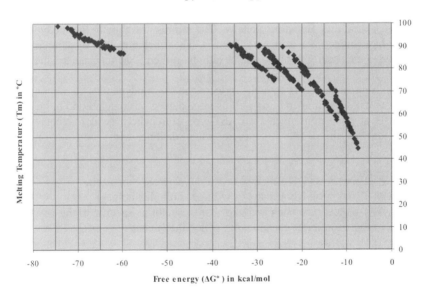

Figure 12-3. A dot plot of HyTher predicted free energy versus predicted melting temperature in 0.1M NaCl for random exact matching oligonucleotides of lengths 10,15,20,25 and 50 (forming clear groups from lower right to upper left).

Appropriate setting of the position-specific scores in a profile allows the raw profile score to encode the caloric values of the oligo's binding free energy. The properties encodes are:

1. A match score is the molar caloric free energy contribution of the matched base and its 5' neighbour, and a portion of the unified model's length-dependent salt concentration penalty.
2. A mismatch score includes the free energy contribution of (a) the matched 5' neighbour and mismatched base (b) the matched 5' neighbour and mismatched base on the opposite strand (c) discount for the NN contribution in the next position.
3. The gap insertion penalty reflects the NN free energy penalty for single base bulges in a duplex.
4. The start of the sequence encodes the unified model's self-complementarity penalties if applicable. Mismatches in this position also encode mismatched end thermodynamics.
5. One extra state at each end of the profile sequence encodes dangling end thermodynamics in the case that the oligo matches to either terminus.

The per-position scores of the profile are added up as nearest neighbour free energy (ΔG) contributions would be in traditional programs, minus the unified models ~1kcal/mol initiation penalty. The nearest neighbour thermodynamics of single base bulges are estimated from[23]. The final raw PSSM score is the real thermodynamic score. For a detailed example of the PSSM mechanism, the reader is referred to[21].

This representation accurately reflects the thermodynamics of oligonucleotide duplexes with dangling ends, as well as interspersed mismatches and bulges. Such a search is advantageous over a BLAST-type search because it overcomes several limitations of the BLAST heuristic when dealing with short oligonucleotide sequences:

1. BLAST may miss DNA matches with interspersed mismatches
2. BLAST may miss DNA matches with gaps (duplex bulges)
3. BLAST match, mismatch and gap scores cannot be specific to the DNA context (NN model of thermodynamics) in which they occur

Due to these limitations, oligonucleotides with no apparent secondary binding with BLAST may in fact have some, when searched using profiles (increased sensitivity). Also, candidates rejected due to a percentage similarity cutoff exceeded in BLAST, may in fact not bind strongly to those sites when the NN thermodynamics are calculated (improved precision).

12.5.4 Special Optimizations for Sequencing Primers

Osprey can be controlled using a contig information file from the Staden sequence assembly package[24] to which the user can add control information. The Staden data is generated using the "Show relationships" option in gap4, the output of which is saved to disk as in Table 12-3.

```
===============================================================
Mon 12 May 14:31:09 2003: Database information
---------------------------------------------------------------
Database size        8000     Max reading length    4096
No. Readings          248     No. Contigs             36
No. Annotations        61     No. Templates          270
No. Clones              1     No. Vectors              1
Total contig length 77917     Average length      2164.4
Total characters in readings                      177474
Average reading characters per consensus character  2.28
Average used length of reading                     715.62
Current maximum consensus length is 168682


===============================================================
Mon 12 May 14:31:20 2003: show relationships
---------------------------------------------------------------
                    238    7039    429    198        0
  CONTIG LINES
  CONTIG        NUMBER  LENGTH         ENDS
                                   LEFT   RIGHT
                    2    1536          5      16
  GEL LINES
  NAME                  NUMBER POSITION LENGTH   NEIGHBOURS
  NO 5PRIME                                     LEFT   RIGHT
  Z02C04FP_B07              5       1    -323      0       6
  Z02C04RP_B07              6      32     301      5     204
  b07w07.07              204     272    -390      6     212
  z02e02Rp_b07           212     473    -521    204      15
  Z01C12FP_B07            15     543     853    212     100
  Z01B09FP_B07           100     720     757     15     136
  Z01B10RP_B07           136     813     929    100     101
  z02h07Fp_b07           101     975    -350    136      81
  Z01G01FP_B07            81    1073     814    101     147
  Z01E09FP_B07           147    1104     857     81      16
  Z01C12RP_B07            16    1536    -886    147       0


  CONTIG LINES
  CONTIG        NUMBER  LENGTH         ENDS
  IGNORE                                       LEFT   RIGHT
```

Table 12-3. Section of a Staden plain text assembly information file with Osprey directive inserted. If an assembly information file is given, the program calculates default primers only for those contiguous sequences (contigs) that are not marked IGNORE on the left hand side of the line denoting left and right neighbour columns. In walking mode, "NO 5PRIME" in that line location will cause the 5' primer not to be calculated. Bold characters are inserted for syntax emphasis only. Similarly, a "NO 3PRIME" statement could be inserted in the same place if no 3' oligo is desired.

The sequence itself can either be represented in Staden's "Strand coverage" view or as a plain FastA sequence file (in case an assembly engine other than gap4 is used). When in the sequence polishing (disambiguation and quality control) mode, regions with ambiguities and single-strand coverage are broken down into per-strand problem spot lists. The positive strand list is analyzed first, starting with the furthest upstream problem location. The candidate primers are checked from furthest downstream to further upstream with the expectation that early successful candidates may allow a single sequencing reaction to also cover other problem locations slightly downstream with good quality sequence, as shown in Figure 12-1 earlier. All problems downstream on the same strand as the priming site, within the expected sequencing read length, are assumed to be resolved by the sequencing reaction, and therefore are removed from the problem list. Single strandedness problems on both strands are checked, followed by unresolved ambiguities. This can reduce the number of primers required since the strand of sequencing for disambiguation is not usually essential. If a plain sequence file is provided instead of a Staden assembly file, only ambiguities can be resolved.

The calculation of linking primers presents a special case of disambiguation, where the 5' and 3' ends of contigs are marked as problem locations. In the linking mode, the assembly information can be marked up (see Table 12-3) to exclude the ends of some contigs. For example, the *sp6* or *t7* ends of cosmids should not be included in the list of walking candidates, as the sequence generated would only contain vector information. Contig ends can also be marked up to request the second or third primer candidate in case the first walking primer has been applied unsuccessfully. Walking mode also requires a minimum assembly overlap length parameter (see Section 12.2.2.1) as an anchor to ensure that the resulting sequence read starts within the contig on which it was calculated.

IGNORE	No primers are calculated for this contiguous sequence
NO 5PRIME	No 5' end extension primer is calculated for this contiguous sequence
NO 3PRIME	No 3' end extension primer is calculated for this contiguous sequence
N-5PRIME	Chooses the *N*'th best 5' end extension primer (e.g. if the first one Osprey suggested failed in the lab)
N-3PRIME	Chooses the *N*'th best 3' end extension primer (can be combined with the previous option with comma separator)

Table 12-4. *Valid Osprey oligonucleotide selection directives to insert into the assembly file as per Table 12-3.*

12.6 Conclusions

Osprey is a software package for oligonucleotide design that uses the unified NN model and simplifies the user's parameterization. Osprey automatically sets many criteria and threshold to quickly find optimal oligos. It filters repetitive sequences and track best sub-optimal oligos for troublesome sequences. It also employs novel methods for the minimization of off-target binding.

The design of oligonucleotides involves the selection of a target sequence range and filtering of candidate substrings within that range based on many common criteria. The selection of criteria thresholds is dependent on the G+C% content of the sequences involved and the length of the oligonucleotide. Formulae can be applied to estimate optimal and threshold values. Methods for the calculation of the criteria values can vary in their complexity and accuracy, with those taking thermodynamics into account being the preferred choice. Various lab techniques such as PCR product amplification also require the application of additional criteria. The reader should now be familiar enough with criteria evaluation and computations methods to set parameters for, and to judge the competency of, oligo design programs.

References

1. Watson, J.D., Crick FH: Molecular structure of nucleic acids; a structure for deoxyribose nucleic acid. *Nature* 1953;171:737-738.
2. Broude, N.E.: Stem-loop oligonucleotides: a robust tool for molecular biology and biotechnology. *Trends Biotechnol.* 2002;20:249-256.
3. Pozhitkov, A.E., Tautz, D.: An algorithm and program for finding sequence specific oligonucleotide probes for species identification. *BMC Bioinformatics* 2002;3:9.
4. Khvorova, A., Reynolds, A., Jayasena, S.D.: Functional siRNAs and miRNAs exhibit strand bias. *Cell* 2003;115:209-216.
5. Silva, J.M., Sachidanandam, R., Hannon, G.J.: Free energy lights the path toward more effective RNAi. *Nat. Genet.* 2003;35:303-305.
6. Amarzguioui, M., Prydz, H.: An algorithm for selection of functional siRNA sequences. *Biochem. Biophys. Res. Commun.* 2004;316:1050-1058.
7. Reynolds, A., Leake, D., Boese, Q., Scaringe, S., Marshall, W.S., Khvorova, A.: Rational siRNA design for RNA interference. *Nat. Biotechnol.* 2004;22:326-30.
8. Wallace, R.B., Shaffer, J., Murphy, R.F., Bonner, J., Hirose, T., Itakura, K.: Hybridization of synthetic oligodeoxyribonucleotides to phi chi 174 DNA: the effect of single base pair mismatch. *Nucleic Acids Res.* 1979;6:3543-57.
9. Rychlik, W., Rhoads, R.E.: A computer program for choosing optimal oligonucleotides for filter hybridization, sequencing and in vitro amplification of DNA. *Nucleic Acids Res.* 1989;17:8543-51.
10. Devoe, H., Tinoco, I. Jr: The stability of helical polynucleotides: base contributions. *J. Mol. Biol.* 1962;4:500-517.
11. SantaLucia, J.: A unified view of polymer, dumbbell, and oligonucleotide DNA nearest-neighbor thermodynamics. *Proc. Natl. Acad. Sci.* USA 1998;95:1460-1465.
12. Rice, P., Longden, I., Bleasby, A.: EMBOSS: The European Molecular Biology Open Software Suite. *Trends in Genetics* 2000;16:276-277.
13. Rozen, S., Skaletsky, H.: Primer3 on the WWW for general users and for biologist programmers; in Krawetz S, Misener S (eds): Bioinformatics Methods and Protocols: Methods in Molecular Biology. Humana Press, Totowa, NJ, 2000, pp 365-386.
14. Breslauer, K.J., Frank, R., Blocker, H., Marky, L.A.: Predicting DNA duplex stability from the base sequence. *Proc. Natl. Acad. Sci.* USA 1986;83:3746-50.
15. Zuker, M.: Mfold web server for nucleic acid folding and hybridization prediction. *Nucleic Acids Res.* 2003;31:3406-3415.
16. Bedell, J.A., Korf, I., Gish, W.: MaskerAid: a performance enhancement to RepeatMasker. *Bioinformatics.* 2000;16:1040-1041.
17. Volfovsky, N., Haas, B.J., Salzberg, S.L.: A clustering method for repeat analysis in DNA sequences. *Genome Biol.* 2001;2:RESEARCH0002 (epub).
18. Altschul, S.F., Madden, T.L., Schaffer, A.A., Zhang, J., Zhang, Z., Miller, W., Lipman, D.J.: Gapped BLAST and PSI-BLAST: a new generation of protein

database search programs. *Nucleic Acids Res.* 1997;25:3389–3402.

19. Smith, T.F., Waterman, M.S.: Identification of common molecular subsequences. *J. Mol. Biol.* 1981;147:195-197.

20. Pearson, W.R.: Flexible sequence similarity searching with the FASTA3 program package. *Methods Mol. Biol.* 2000;132:185-219.

21. Gordon, P.M.K., Sensen, C.W.: Osprey: a comprehensive tool employing novel methods for the design of oligonucleotides for DNA sequencing and microarrays. *Nucleic Acids Res.* 2004;32:e133 (epub).

22. Gribskov, M., McLachlan, A., Eisenberg, D.: Profile analysis: Detection of distantly related proteins. *Proc. Natl. Acad. Sci.* USA 1987;84:4355-4358.

23. LeBlanc, D.A., Morden, K.M.: Thermodynamic characterization of deoxyribooligonucleotide duplexes containing bulges. *Biochemistry* 1991;30:4042-7.

24. Staden, R.: The Staden sequence analysis package. *Mol. Biotechnol.* 1996;5:233-241.

Index

ISBN: 1-933255-07-2

DNA ARRAY IMAGE ANALYSIS NUTS & BOLTS
SECOND EDITION

Edited by
Gerda Kamberova, Ph.D. and
George Kamberov, Ph.D.

The book offers the basics of machine vision and image analysis for biologists, "how-to" approach to microarray image analysis, and practical tips for both the novice and the advanced user in this field.

In this book you will learn:
- Image analysis concepts
- Microarray scanners - the pros and cons
- Background corrections
- Correcting for systematic errors in your image analysis
- Microarray image analysis in minutes with software tools
- New developments in microarray image technology

"This book...will be useful reference not only for computer and biology scientists, but for anyone using or interested in microarray technology."
Ming Dong, Professor,
Wayne State University

"The book DNA Array Image Analysis is the most comprehensive book that I have read, dealing with microarray analysis topic. This book gave me a very nice perspective on statistical approaches for microarrays. I recommend it to all colleagues, especially for those who are running microarray facilities."
Greg Khitrov, Manager of Gene Array Resource Center,
The Rockefeller University, NY, NY USA
Posted at www.amazon.com

DNA Press™

THE NUTS & BOLTS SERIES

MICROARRAYS
METHODS
AND APPLICATIONS
SECOND EDITION

Gary Hardiman, Ph.D.
Editor

ISBN 1-933225-15-3

MICROARRAYS METHODS AND APPLICATIONS
SECOND EDITION

Gary Hardiman, Ph.D.
Editor

This book is a "how to" manual for the laboratory researcher who is using or plans employing DNA or protein micorarrays. It is packed with:
• Useful protocols for various microarray technologies
• Statistical experimental design concepts
• Microarray image analysis tips
• Microarray data analysis recommendations
• Analysis of microarray automation and software
• Applications in many research fields including pharmacology

"An easily accessible and authoritative guide to microarray technologies, for the student and the laboratory manager alike. With focus on spotted cDNA arrays, it deals adequately not only with the microarray chip and its hybridization targets, but also with the related issues such as laboratory automation and data analysis. The scope ranges from scientific definitions, to extensive and valuable protocols; and comprehensive application examples from fields as diverse as agricultural studies to more conventional pharmacology."
Markus Heilig, M.D., Ph.D.
Associate Professor, Director NEUROTEC,
Karolinska Inst., Huddinge University Hospital, Sweden

"This book will be of great use for both post-graduate students and scientists in the field. The chapters are very detailed and full of practical and inside tips to produce high quality microarrays."
Paul Van Hummelen, Ph.D.,
Research Manager,
VIB Microarray Core Facility, Belgium
www.microarrays.be

DNA Press™

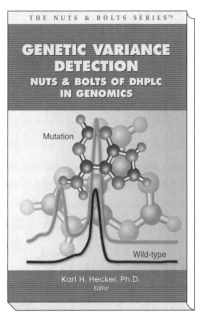

ISBN: 09664027-7-4

GENETIC VARIANCE DETECTION:
Nuts & Bolts of DHPLC In Genomics

Karl H. Hecker, Ph.D.
Editor

This book offers insights into genetic variance detection with DHPLC. Implementation of the technology is shown for gene specific applications with projects ranging in size from small and focused experiments to comprehensive, multi-center studies. The book also covers novel and innovative applications of the DHPLC technology and demonstrates the potential scope of a highly versatile and powerful technology for genetic variance detection.

In this book you will learn:
- The fundamentals of DHPLC for genetic variance detection
- PCR requirements for successful variance detection
- Rational experimental design by example
- Case studies for gene specific variance detection
- Scope and potential of DNA chromatography

" In this book Dr. Karl Hecker systematically covers all aspects and major applications of DHPCL technology for genetic variance detection. The chapters in this unique book will serve as a resource for all those interested in this important technology and serve a a catalyst to future applications and technology improvements."

E. Holinski-Feder, Dr. med. Dipl. chem.
Medical Genetics Centre, Munich (MGZ), Germany

DNA Press™

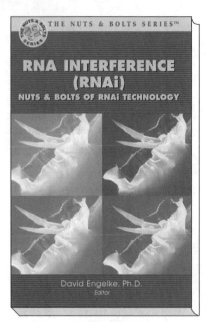

THE NUTS & BOLTS SERIES™

RNA INTERFERENCE
(RNAi)
NUTS & BOLTS OF RNAi TECHNOLOGY

David Engelke, Ph.D.
Editor

ISBN: 09664027-8-2

RNA INTERFERENCE (RNAi)
Nuts & Bolts of RNAi Technology

David R. Engelke, Ph.D.
Editor

This book offers protocols and advice for the use of RNA interference for knocking down expression of target genes in a number of different eukaryotes. Varied applications in mammalian systems are explored in depth by authors currently working at the leading edge of RNA interference methodologies. Chapters will include:

- RNAi in plants
- RNAi in *Drosophila*
- RNAi in *C. elegans*
- RNAi in mammalian systems
- Design, synthesis and preparation of synthetic RNAi
- Expression of RNAi from recombinant DNA and viral vectors
- High-throughput RNAi applications
- Emerging alternative RNAi methods
- RNAi in mice

"This book offers a comprehensive guide to RNAi experimentation and provides in depth coverage of the application of this technology in plants, flies and mammalian cells. The chapters reflect recent developments in this rapidly growing field. Written in clear concise language, "RNA Interference (RNAi) ~ Nuts and Bolts of RNAi Technology" serves as an ideal resource for undergraduates and experienced research scientists."

Gary Hardiman, Ph.D.
Director BioMedical Genomics Microarray Facility
Assistant Professor, Department of Medicine
University of California San Diego

DNA Press™